Frederic R. Sturgis

The Student's Manual of Venereal Diseases

being the university lectures delivered at Charity Hospital, B.I., during the Winter

session of 1879-'80. Fifth Edition

Frederic R. Sturgis

The Student's Manual of Venereal Diseases
*being the university lectures delivered at Charity Hospital, B.I., during the Winter session of
1879-'80. Fifth Edition*

ISBN/EAN: 9783337255664

Printed in Europe, USA, Canada, Australia, Japan

Cover: Foto ©berggeist007 / pixelio.de

More available books at **www.hansebooks.com**

THE

STUDENT'S MANUAL

OF

VENEREAL DISEASES,

BEING THE

*University Lectures Delivered at Charity Hospital, B.I.,
during the Winter Session of 1879-'80.*

BY

F. R. STURGIS, M.D.,

CLINICAL LECTURER ON VENEREAL DISEASES IN THE MEDICAL DEPARTMENT OF THE
UNIVERSITY OF THE CITY OF NEW YORK ; ONE OF THE VISITING SURGEONS
TO CHARITY HOSPITAL, B. I., DEPARTMENT OF VENEREAL AND SKIN ;
ONE OF THE VISITING SURGEONS TO THE NEW YORK
DISPENSARY, DEPARTMENT OF VENEREAL AND
SKIN ; MEMBER OF THE NEW YORK
DERMATOLOGICAL SOCIETY,
ETC., ETC., ETC.

FIFTH EDITION

NEW YORK:
G. P. PUTNAM'S SONS,
27 AND 29 WEST 23D ST.
1887.

𝔗𝔬

THE STUDENTS

OF THE

MEDICAL DEPARTMENT OF THE UNIVERSITY OF THE
CITY OF NEW YORK,

THIS MANUAL ON VENEREAL DISEASES

𝔌𝔰 𝔍𝔫𝔰𝔠𝔯𝔦𝔟𝔢𝔡,

BY THEIR OBLIGED FRIEND AND TEACHER,

THE AUTHOR.

PREFACE.

It has been said, with much truth, that books are read in inverse proportion to their length, and in preparing this Manual I have steadily kept the question of length in view.

Written for students of medicine, it has been my aim to make the book concise, and at the same time practical. I have, therefore, as far as possible, eschewed all mooted points in Venereal medicine, and confine myself to giving a careful, and at the same time condensed, description of the commoner forms of Venereal diseases which will fall to the lot of the average young practitioner to treat, together with the most appropriate remedies.

How well I have accomplished my task remains for others than myself to say. I trust, however, that it will satisfy a want which, from my experience as a lecturer in this branch, I know exists, and with this hope I send the little Manual into the world to take its chances.

16 West Thirty-second Street,
New York City.

CONTENTS.

CONTENTS.

VENEREAL DISEASES.

LECTURE I.

SIMPLE VENEREAL ULCER AND ITS COMPLICATIONS.
—THE CHANCROID.

GENTLEMEN :—Before calling your special attention to the cases which I have brought from the wards
for the purposes of illustration, it may not be inapt
to define what is meant by venereal diseases, and to
set before you the principal groups into which they
are divided.

Speaking broadly, venereal diseases are those due
to, and originating in, sexual contact, and, although
many forms of these diseases are transmitted without
any sexual contact, as I shall show you further on, the
name may, for convenience' sake, stand. They are at
present divided into three principal groups or divi-
sions: *Gonorrhœa, Chancroid*, and *Syphilis*. Each
is distinct and separate one from the other, having
nothing in common with each other, although they
may all be present upon the same person at the same

I

time, and possessed of certain characteristics which are more or less peculiar to themselves.

Of these three diseases, only the last one, syphilis, is constitutional; the other two, gonorrhœa and chancroid, are local. Remember, then, *gonorrhœa and chancroid are local. Syphilis is not; it infects the entire system.*

In the lectures which it was my pleasant duty to give before you in the spring course, I dwelt at length upon gonorrhœa; less so upon chancroid and syphilis, and this with a purpose.

Here, in the amphitheatre of the hospital, I can bring you face to face with cases of syphilis; you can have better facilities for examining the cases for yourselves than at the college, and if you expect to learn, you must handle and examine the cases yourselves; no one else can do so for you, and the first case I present for your consideration is a case of chancroid in a male subject.

The history, I regret to say, is imperfect, no uncommon occurrence in cases coming into this hospital; but from what I can glean from him and the record book, his sore, which you see is a pretty large one, came on two or three days after coitus, and was at first quite small. Here is a point to which I wish you to attend, one of the most important upon which to base your diagnosis of a *chancroid*, and equally noteworthy as a differential mark between this lesion and the first manifestation of syphilis—what is commonly known as *chancre. The sore came on two or three days after coitus;* in other words, but a short time elapsed between the infecting connection and the resulting ulcer.   The effect was almost immediate.

When we come to treat of syphilis we shall find that this is no longer true ; an appreciable interval elapses between cause and effect ; what is technically called the period of incubation.

Chancroids, then, have at the most a *very short* period of incubation, sometimes *none* at all, and this depends much upon the manner in which the poison, or virus, so-called, is deposited beneath the mucous membranes.   If in coitus the membrane is abraded or torn, the chancroidal action begins at once, while, on the other hand, it is delayed if the matter is deposited in a fold of mucous membrane, or in a follicle ; but even then the delay is one usually of only thirty-six to forty-eight hours.

Another circumstance in the case is worthy of remark ; the ulcer has *increased in size*, at first he says it was quite *small.*   This denotes, in chancroids, a tendency to spread and become larger instead of smaller, a tendency due to the destructive character of the poison.   Let me say here a word or two about this virus or poison.   It holds, in venereal parlance, much the same position that the letters x, y, and z do in algebra, it is an unknown quantity.   No one has yet demonstrated the existence of the virus of chancroid, or of syphilis, except by the results ; and on the principle that what is non-apparent is non-existent, some writers entirely deny the presence of a virus, and claim these two diseases as due to some other cause. Be that as it may, the term is one of great convenience, and it would be difficult to find a good substitute. I shall therefore, in these lectures, retain it, and I beg you will remember that it means an indefinite something endowed with certain properties, which varies in these two diseases, and produces different results.

To return to our chancroid. Two points we have brought out, and mark them well. 1st, *a period of incubation, at the most very short, sometimes absent;* and, 2d, *a tendency to destructive action.* Let us now examine the sore and see what else we find. We notice *one* rather large ulcer of *irregular* shape, *uneven* floor, a moderately *copious, purulent* secretion (this has been somewhat modified by treatment), and on putting the ulceration on the stretch we observe that it extends beyond the *apparent* edges of the sore. I repeat *apparent edges,* because this peculiarity has a decided bearing upon treatment. Chancroids frequently burrow, going along faster below than they do above, hence the external aspect of the sore is no necessary index of its real area; the edges of the ulcer are *undermined,* and if, in the treatment, you decide to destroy the chancroid by caustics, *convey the destructive agent beneath the edges and beyond the apparent limits of the sore, even into sound tissues.*

The number and shape of the ulcers are the next

points which invite discussion. In this subject there happens to be only one, but such is not always the case, as witness this second man. Here we find three chancroids of various sizes. This multiplicity may be produced in one of two ways: either as original foci of infection, or by inoculation.

Note, therefore, *that the chancroid* is capable of self-propagation upon the person having it, and also upon others to whom the poison may be conveyed.*

It is eminently *contagious* and *auto-inoculable.* I shall call your attention again to this point when I come to speak of the initial lesion of syphilis (chancre). In shape the sore is irregular, owing partly to the seat, on the inner layer of the prepuce and the fossa glandis, and partly to the natural tendency chancroids have of spreading irregularly and sending out shoots, but there is often another reason. Several chancroids may be seated close to each other, and, by destroying the intervening sound tissue, present to view an ulceration with irregular scalloped edges.

The secretion, as has been already said, is *copious* and *purulent,* caused by the *destruction* of tissue, and pus, as you know, is *débris.*

Have we explained all the noteworthy characters presented by this chancroid ? By no means, for upon handling it we are struck by the fact that though the ulcer is large and angry-looking, the tissues upon which it is seated are perfectly supple and soft. And here let me give you a word of warning as to the use of the word *soft*, which has proved a fruitful cause of misunderstanding. Better expunge the word from your venereal vocabulary and call the chancroid a *simple venereal ulcer*, as contradistinguished from the initial lesion of syphilis (chancre), which is termed the *specific venereal ulcer*. In the discussions which in former days have been had upon the nature of these two ulcers, it was stated and generally believed that no soft sore, *i. e.*, one which had no induration at the base, was ever followed by syphilitic manifestations. This belief is now proved to be erroneous, and sores devoid of indurated bases have been the precursors of secondary symptoms—in other words, *the initial lesion of syphilis may be soft.* The importance of this you will see later on. If this be true, the inapplicability of the term to a chancroid is apparent, for a *chancroid is never followed by general manifestations, the initial lesion always is,* and this I shall speak of more fully later on. Do not therefore call the chancroid the *soft* venereal ulcer, but *simple* venereal ulcer, if you do not wish to use the word chancroid.

Now, to come back to our chancroid, which has been waiting for us. We find *no* induration at the base ; the tissue upon which it is seated is perfectly supple and yields readily to pressure, in a manner entirely different from what it does in this third patient, who is the subject of an initial lesion and beneath whose sore, on palpation, you can discover a gristly hard substance, the nature of which you will learn more about by and by. We have, then, discovered another trait of a chancroid, to wit, an absence of indurated base ; but remember this loses some of its diagnostic importance, from the fact that the initial lesion (chancre) sometimes presents the same peculiarity.

Before passing to the next point let me sketch upon the blackboard the salient features of a chancroid, such as we have discovered upon the cases examined to-day. They are these :

Absence, or at most a very short period of incubation.

Tendency to spread irregularly in size and depth.

Tendency to undermining of the walls of the ulcer.

Copious purulent secretion.

Contagious and auto-inoculable character of the pus, thus producing multiple sores.

Absence of induration of the base of the ulcer.

Thus far we have studied the chancroid in its simplest form. We will now consider the complications most liable to occur with this disease.

The first and the one most intimately associated with the chancroid, is the bubo or swelling of the glands, usually those of the groin. This is due to two causes : the first being sympathetic inflammation; the second and most serious, absorption of the chancroidal matter from the ulcer, by the lymphatics.

The two subjects I bring before you illustrate these points beautifully. This first case has, as you see, a large, indolent, brawny swelling in his right groin; and upon his penis he still bears a chancroid, but in the stage of repair. Number two also has a chancroid seated close to and invading the frænum, but in his groin we find a different condition of things to what we did in number one. Here we find an open ulceration presenting an uneven, grayish floor, everted and undermined edges, and secreting an abundant amount of pus, recalling to mind the characteristics of the chancroid already presented to you. Indeed, you would be right to call it a chancroid, for such it is ; caused by the absorption of the chancroidal matter through the chain of lymphatics, and deposited in the nearest glands (in this case the inguinal), there to produce a similar condition of affairs to what obtains in the original ulcer. In other words, you have here a typical chancroidal bubo, pure and simple. These two, then, represent the varieties of bubo found with a chancroid ; the first one, a bubo from sympathy, which frequently does not suppurate, and if it does, furnishes *laudable healthy pus ; the second one, the true chancroidal bubo, due to absorption of the matter from the ulcer, which invariably ulcerates and presents subsequently the appearance of a chancroid—indeed, is a chancroid.*

There are two other points to which I wish to call your attention : the diffused brawniness of the surrounding tissues in both cases, and the side of the body upon which the buboes are seated.

The glands themselves do not seem to be the only parts affected; the circumglandular tissue is involved as well, presenting a thickened doughy mass in which the glands can be indistinctly felt. Note this well, I pray, for when you come to handle cases of syphilis you will find a very opposite condition of things; the glands will not be fused together nor with adjacent tissue, but they will be *distinct, well marked,* and *indurated.*

The other point is this: in the second case the bubo is seated upon the groin opposite to the side of the penis upon which the chancroid is; in number one it is upon the same side. The cause is the position of the ulcer. Deduce then the following rule: *Buboes are usually seated upon the same side of the body as the ulcer which causes them, except when it (the ulcer) is seated upon the frænum, when they will be frequently found upon the opposite side. The same is true when the chancroid in the female is seated upon the "fourchette," and this is due to the decussation of the lymphatics at these two points.*

In all the cases I have shown you, the lesion has been seated upon the mucous membrane of the genitals. This is its usual seat, but it may be met with upon the skin of various parts of the body, such as the face, the fingers, and—as I have seen in one case—in the throat. Such places are not common seats of the

1*

chancroid, so you may always suspect the nature of
a sore when located on the parts I have just mentioned; it is much more likely to be *syphilitic;* at
any rate, always bear that point in mind.

The course of a chancroid is always *destructive,*
and if not properly treated may result in severe disfigurement and loss of tissue.

This is especially the case when the chancroid is
seated upon the frænum, or in the urethra just within
the meatus urinarius.   In the former place, perforation and destruction of the frænum is to be looked
for; and what will perhaps surprise you, is a greater
loss of tissue than you had at first counted upon, for
here, particularly, the burrowing tendency of the
chancroid is shown, and long before the frænum is
ulcerated through, the sore has attained to large di-
mensions.   In the latter place (the urethra) the sore
extends rapidly, is difficult of treatment from its
comparative inaccessibility, and upon cicatrization
produces partial stenosis of the meatus, requiring subsequent surgical treatment.

As I have already stated, these ulcers have a tendency to spread, and from their facility of auto-inoculation, to multiply; hence the treatment, to be effective, must be *prompt* and *thorough.*   Under proper
care the copious purulent secretion is diminished,
the gray floor disappears, granulations spring up over
the surface of the sore, and the undermined edges fill
up level with the walls of the ulcer.   But, bear this

point in mind : a *chancroid is dangerous up to the very moment of its complete cicatrization ; no matter how superficial or simple it may look, do not remit the thoroughness of your treatment until cicatrization is complete.* I have seen chancroids *almost* well relapse (without a fresh infection) from just that want of care, the slight discharge remaining being sufficient to re-inoculate the almost cicatrized sore.

Phagedena is another and perhaps the worst accident which can attack a chancroid, and when it becomes serpiginous—that is, when it extends in one direction while healing in another—may last for a long time (several years) and seem well-nigh hopeless of cure. It fortunately is not common in this section of the country at least, and occurs in those persons whose health is broken down from alcoholic excesses or constitutional debility, such as scrofula and the like. Remember, that it is due to *constitutional, not local causes*, and to combat it successfully you must take your measures accordingly. This grave accident attacks not only the chancroid itself, but the chancroidal bubo, lasts for an indefinite time, and will put you to your trumps to cure.

Before passing on to the consideration of treatment, there are other complications to which I wish to direct your attention, to wit : phimosis occurring with chancroid, and chancroids of the anus. You already know that the first of these complications occurs with syphilis and gonorrhœa, as well as with chancroids, and it is important for you to be able to know which one of these diseases lurks behind the constricted foreskin, not only for the diagnosis, but for treatment.   In cases of clap and chancroid there is a copious purulent secretion from beneath the prepuce ; but in *gonorrhœa* this matter *is not auto-inoculable*, while in *chancroid it is*.   With a chancroid the penis is much more painful, œdematous, and swollen, and the lymphatics on the dorsum penis are more apt to be inflamed and tender than is the case in gonorrhœa ; *but the crucial test is auto-inoculation*.   If the hidden ulcer be an initial lesion, the secretion is *very scanty*, if indeed there be *any ;* the prepuce is *hard* and *indurated*, instead of being *œdematous* and *doughy*, and the *secretion is not auto-inoculable.*

Chancroids of the anus are, in the male subject, very rare indeed ; and where you find them, always suspect sodomy, and I believe you will seldom be wrong.   The same is still more true as regards the initial lesion of syphilis (chancre).   But with women it is different.   With them anal and rectal chancroids are not rare, and their presence does not imply *Venus præpostera.*   The secretion from the chancroids of the female genitals naturally flows over the perineum and anus ; very few feminine ani but are abraded ; auto-inoculation occurs, and the thing is done, and a very nasty thing, too.   The ulcer extends in all directions, eats through and neutralizes the action of the sphincter ani, producing incontinence of the bow-

els ; burrows up into the rectum ; is continually irritated by retained fecal matter ; is extremely difficult
to heal, and, when it finally does, nearly always
leaves a stricture of the rectum behind it; and if to
that you add phagedena, a not infrequent complication in broken-down harlots, the picture is a pretty
dismal one.

Chancroids of the female genitals differ in no essential respect from those of the male in appearance or
course.   Their usual seat is at the vulva and introitus
vaginæ ; they are next most frequent on the cervix
uteri, and are very rarely met with in the vagina between these points.

Buboes in women are not so common as in men,
excepting when the chancroid is seated at the "fourchette," when they follow the same course of action
as that already detailed in the early part of this lecture.

LECTURE II.

TREATMENT OF THE CHANCROID.

At our last meeting we went over the description of the chancroid and the complications which are its most frequent concomitants, reserving the question of treatment to a lecture by itself. This, then, will form the subject of to-day's lesson, and, at the outset, I want to impress upon your minds the two cardinal points of treatment, which are, *first, the arrest of the virulent and destructive character of the ulcer; second, cleanliness.*

First, then, as to the arrest of the virulent and destructive character of the ulcer. This is done either by the actual cautery or other caustics, in severe cases, and by alterative applications in light ones. Of the first division of remedies the *white iron*, or the *galvano-cautery*, takes the front rank as a destructive agent; next to that comes the *strong sulphuric acid;* third, chemically pure *nitric acid;* and fourth, pure *carbolic acid.* A neat way of using the sulphuric acid is the method known as Ricord's carbo-sulphuric paste, which is made by taking a small quantity of finely-powdered *willow* charcoal, adding, drop by drop, enough of the acid to make a paste of the consistence of thick cream. This is put on with a porcelain or glass spatula, *taking care (remember the undermined edges) to carry the agent into sound tissue both underneath and on the surface of the edges of the chancroid.* Nitric or carbolic acid may be used in the same way. The advantage of this method is, that besides destroying the virulent ulcer, it makes a firm dressing by the drying of the charcoal on evaporation of the acid, which, on dropping off at the end of several days, reveals the chancroid almost if not entirely healed up. If you prefer to use the acids in a fluid form, then some subsequent dressing must be used, and of all dressings I infinitely prefer the dry to the wet. One of the best preparations is iodoform *finely* powdered, either alone or in combination, thus:

> ℞. Pulv. iodoformi.................. 1 part.
> Lycopodii..................... 2 parts.
> Triturate well—apply locally.

The lycopodium has, probably, only a mechanical action, but it absorbs fluid very readily, while the iodoform acts as a local stimulant and alterative. Another good prescription is,

> ℞. Pulv. iodoform.
> Pulv. ac. tannic. p. œ.
> Triturate and use locally.

This is more astringent than the other.

No. 3. is useful when the ulcer looks flabby and indolent.

℞. Pulv. iodoform............. ..... ℨ i.
Zinci sulphat..................... gr. v.
Pulv. ac. tannic.................. ℨ i.
M. Triturate.   For local use.

One serious objection to iodoform in private practice is the strong and pungent smell which it has. Many attempts have been made to overcome this, and Dr. Bronson of this city speaks highly of combining the iodoform with some essential oil, such as peppermint, rosemary, and the like, which he claims overcomes the odor without interfering with the alterative action of the drug.*

Should you from any cause decide to use a wet in preference to a dry dressing, you will find the formulæ which I give below as good as any you can use :

℞. Ac. carbol. cryst............. ℨ i.–℥ ij.
Aquæ..................... ℥ viij.
M.
Or—
℞. Zinc. sulphat................. gr. v.–xx.
Aquæ..................... ℥ ij.

---

* The following is the formula :
℞. Iodoformi pulv........................... ℨ i.
Mucilag. acac.,
Glycerinæ ....āā gtt. x.
Ol. menth. pip. (seu neroli seu caryophylli)..... gtt. i.
Misce.

This latter application is an excellent dressing where the ulcer looks flabby and indolent. The strength of 20 grs. to ℥ ij. should only be used when the ulcer is unattended with inflammation ; if there be any, the weaker solution is better.

Another very excellent dressing for chancroids is a weak solution of nitric acid, thus :

℞.　Acidi nitrici c. p.................... ℨ i.
　　Aquæ........................... ℥ viij.
M.

You observe that, in the list I have written for your use, the nitrate of silver does not appear. This may appear strange, for the lunar caustic is the one *par excellence* which is daubed over any suspicious-looking ulcer. But I say to you, *don't use it* if you mean to use a *caustic.* Nitrate of silver is *not*, in the true sense of the term, a *caustic ;* its action is very superficial, inasmuch as it quickly forms, with the albumen of the tissues, an insoluble albuminate of silver, and it cannot destroy deeply or thoroughly as do the sulphuric and nitric acids. *Confine its use, then, to those cases where you desire to stimulate indolent, slowly-healing chancroids ; when you wish to destroy, select some other agent.*

The above rules for treatment are good where the lesion is exposed and accessible, but how shall we act in cases of chancroids concealed either in the urethra or behind a phimosis ? The first object to be attained is to relieve the phimosis, the second to check the extension of the chancroid. For the first point you will find nothing better than freely bathing the genitals in *hot* water (as hot as the patient can bear it, even to the point of faintness) several times daily, and at night wrapping the penis up in the following lotion:

> ℞.   Liquoris plumb. subacetat.,
>   Tinct. opii....................āā ℥ i.
>   Aquæ ad........................ ℥ viij.
> M.   S.—Local use.

In conjunction with the hot bathing, subpreputial injections should be made several times during the day with a solution of carbolic or nitric acid in the following manner : With a Taylor's syringe, which is nothing but a flat-billed syringe, made of hard rubber, throw up hot water between the prepuce and glans penis until the return flow shows no shreds or fibres; then, with the same instrument, inject carefully two syringesful of either of the following solutions :

> ℞.   Ac. carbol. cryst............... ℨ ss.– ℨ i.
>   Aquæ........................ ℥ viij.
> Or—
> ℞.   Ac. nitric. c. p.................... ℨ ss.
>   Aquæ........................ ℥ viij.

taking care that the fluid reaches well back to all portions of the fossa glandis. After this is done, a small dossil of lint or prepared cotton should be lightly placed at the orifice of the prepuce, between it and the glans penis. This plan of procedure should be steadily persevered in until the prepuce can be retracted and the glans penis freely exposed, when the chancroids can be treated as already advised.

Suppose this happy result not attainable, what then must we look for ? It may happen that the swelling and inflammation, instead of subsiding, increases ; the entire organ becomes enormously œdematous and purple, threatening gangrene, and it is evident that a very serious condition of things obtains ; in fact, gangrene *will* rapidly come on unless active measures are adopted to check it.

Sometimes, happily very rarely, the sphacelus attacks a large portion of the penis, causing very serious consequences ; but usually it is confined to narrower limits, and Nature is satisfied when she has relieved the prepuce and re-established the circulation. This she does in the following manner : One or more spots of a purple hue appear upon the swollen prepuce at points corresponding with the imprisoned glans penis beneath ; these spots get darker in color, extend and coalesce, and by becoming gradually thinner admit of the exit of the glans penis through the opening, safe and sound. The redundant and useless foreskin may be subsequently removed by operation. This is the course where everything goes on smoothly and safely, but sometimes active surgical interference becomes requisite. This happens when it is

evident that extensive loss of tissue must supervene before the imprisoned glans penis can be lib-
erated, and here you have to carefully choose between two evils. You must overcome the constriction by cutting through it; but remember what I have
already told you about the contagious character of
the chancroid. *The cut edges of the incision are sure
to become inoculated, hence I advise you not to operate
unless it be done to save your patient from something
worse than an extension of the chancroid.* But if you
have to do it, let me give you one or two hints as to
the way. Carry your director between the prepuce
and the glans penis in the median line * (be careful
not to pass it into the urethra), and then slit the foreskin well up to the fossa glandis ; that will liberate
the glans, and on retracting the prepuce, search for
the chancroids. *Destroy them at once* with one of the
strong caustics already mentioned, *and at the same
time cauterize the cut edges of the wound you have
made.* The subsequent dressing will be similar to
what I have already advised. The " dog's ears " left
by the operation may be subsequently removed by
circumcision, *but not until the chancroids have entirely healed.*

If the chancroid be in the urethra, your tactics
must vary a little. When situated close to the mea-

---

* The incisions are sometimes made upon the two sides instead of the
median line. This variety of incision is better if the foreskin is very
much thickened.

tus, separation of the lips will expose the sore, which may be cauterized and dressed with one of the wet preparations previously mentioned. When beyond reach, upon separation of the lips of the meatus, you must use a weak injection of carbolized * or otherwise medicated fluid, and afterward insert a dossil of lint or cotton, wet with the same solution, within the urethra.

Contraction of the meatus left upon cicatrization of the chancroid may be remedied by slitting the meatus with a bistoury or a meatotome.

Such dressings—indeed all dressings for the treatment of chancroids—should be made three or four times daily, at the least.

When the chancroid is seated at the frænum, threatening perforation, do not wait for the ulcer to eat its way through, but anticipate matters by cutting the frænum yourself. If hemorrhage result from the small artery seated in the frænum, tie, if requisite, but torsion will check bleeding in the majority of cases. You must then treat the chancroid, which will often turn out much larger than you at first supposed, by the rules I have already given you.

As regards the treatment of buboes, the rules are simple and easily laid down. Until the bubo breaks you cannot be certain whether it is a simple or a chancroidal one you have to deal with. Your first

---

* Ac. carbol., ⅛, ¼ gr. to aq. ℥ i.

efforts, therefore, should be to cause absorption ; **if**
the bubo is non-virulent, you are often **successful ;**
but if, on the other hand, the bubo is due to the absorption of matter from the chancroid, you will
find the swelling extend, the bubo rapidly become
softer, and fluctuation more pronounced. *The moment you are sure of fluctuation, open the bubo,* and
this for a twofold reason. Pus, in my experience,
whether due to virulent or non-virulent buboes, is
not absorbed when once it begins to form, and under
these circumstances it is much better evacuated. If
the bubo be a simple one, the moment the pus is let
out the bubo heals up ; if, on the other hand, it be
virulent, the sooner you know it the better for your
patient. But we will suppose that the bubo has not
as yet shown any fluctuation ; what methods shall we
adopt to prevent the formation of pus ? Four ; viz.,
leeches, rest, compression, and the local application
of the tincture of iodine. This latter must be applied *at least* once every day up to the point of vesi-
cation, and as soon as this is accomplished you will
find the employment of the emplastrum plumbi of
service. Compression, if you can persuade your
patient to go to bed, can be best obtained by placing
a bag of small shot, weighing from two to four pounds,
or a brick wrapped in flannel, directly upon the swelling ; if your patient will not keep on his back, use a
layer of compressed sponge and a spica bandage,
which wet as soon as it is applied, when you will get

even and firm compression from the swelling of the sponge. Should your attempts at resolution fail and suppuration threaten, favor it, as far as possible, by the application of poultices.

A word or two with regard to the application of leeches, should you deem them requisite. *Always place them at some distance from, and never on the bubo.* Do not forget this hint, else you will run the risk of inoculating sound tissue from the leech bites, if the bubo should prove to be chancroidal. It is seldom that leeches are of much service, and I should advise you to be chary of their use ; they are not superior to the other methods I have mentioned.

The bubo is now ripe, and is ready for the knife ; how is it to be opened ? I prefer doing so by an incision parallel with the long axis of the body first, and then, if requisite, carry the cut upward and downward in the direction of Poupart's ligament. *Lay sinuses open wherever you find them,* if you hope to make a speedy and permanent cure. After the bubo is thoroughly opened, stanch the bleeding (exposure to the air will suffice in most cases ; if not, use ice-cold compresses), and in cases of simple buboes dress the wound with a weak carbolized lotion applied on cotton or lint. If, however, the bubo be chancroidal, cauterize it first according to the directions already laid down for cauterizing chancroids, and make what subsequent dressings you deem advisable, carefully packing the material *well beneath the undermined edges.*

An honest, free incision is, I believe, nine times in ten the best and quickest way to treat these lesions, but I will mention to you two other methods in vogue. One is by aspiration — *i. e.*, exhausting the bubo of its contents by suction with Dieulafoy's aspirator, or the American modifications of his instrument.

The other is by breaking up the bubo — *i. e.*, churning its contents with a blunt-pointed bistoury — a *small* incision having first been made to admit the entrance of the bistoury. Both of these methods are, of course, only applicable to the non-virulent bubo, and even here I think other methods are preferable.

If *internal* treatment be thought worthy of trial, it must be borne in mind that it is for its tonic effect more than anything else that it is used. Only one remedy is given with the object of checking suppuration, and that is the sulphide of calcium, which may be administered as follows :

  ℞. Calcii sulph...................... $\frac{1}{10}$ gr.
   Mucilag. acac...................... q. s.
 M. In one pill ; of these, 3 to 6 daily.

A very neat way of giving it is as a compressed tablet, made with the sugar of milk, which can be obtained at most apothecaries'.

The tonics most in use are the ferrum sulph. exsic-
cat. gr. i.–iij., or the fer. pulv. gr. i.–ij. in pill form
three times daily ; the sulphate of quinine, or dextro-
quinine, gr. ij.–iij. three times daily ; and ol. morrhuæ
ℨ i. – ℨ ss. in similar doses. Of course you will not
forget nutritious diet and stimulants p. r. n., but I
should advise you to use the latter as little as you
can. *I believe that venereal patients do better, as a
rule, without alcohol.*

There is one other subject in connection with these
diseases which I wish to discuss with you before bring-
ing this lecture to a close, and that is the one of phagedena. It will be sufficient to recall to your mind the
cases of the three women which I showed you a short
time since, where the ulceration had crept over the
nates and down the thighs, up the abdomen and along
the groins, breaking down the recto-vaginal wall and
destroying the labia vulvæ, to impress upon you the
necessity of a vigorous treatment. Remember what
I said to you in a preceding lecture about phagedena,
that it was *due to constitutional, not local causes,* and
this will be the key-note of your treatment ; *although
not to the exclusion of local remedies, your main
reliance must be upon internal and constitutional
measures.* Foremost in this latter class stands the
potassio-tartrate of iron, which Ricord called the
" born enemy of phagedena," and which he is in the
habit of applying both topically and by the mouth.
Thus,

2

R.   Ferri et potas. tart................. ℥ i.
      Aquæ........................... ℥ vi.
M.
S.—Internally, in teaspoonful doses, thrice daily; also for local application p. r. n.

A strongly carbolized lotion will oftentimes be of service as a dressing in phagedenic chancroids, viz. :

R.   Ac. carbol. cryst............... ℨ ij.–v.
      Aquæ........................ Oi.
M.   S.—Locally.

By far the most frequent cause of phagedena is that condition of the system known as " chronic alcoholism," and which it should be your aim to relieve as far as possible.   In such cases you will find the following prescription a serviceable one :

R.   Ol. morrh.................... ℥ ss.
      Dil. phosph. ac............... ℳ x.–xxx.
In one dose.
S.—Three times daily, or oftener if necessary.

This seems to act by toning up the depressed nervous system of chronic drunkards, and giving the body a chance of combating the disease.

Other tonics which are suitable in such cases are those which I have previously mentioned.

Among the local dressings, the potassio-tartrate of iron and the carbolic acid are the best, but I wish to say a few words about the *destruction* of a phagedenic chancroid. The only agents which are of any real value for that purpose are the hot iron and the galvano-cautery ; the other corrosive agents I have previously mentioned are of little use. In applying either of these agents remember to have the heat *white*, not *red*, for two reasons : first, because it is more effective ; second, because it is less painful. Remember also to carry the destruction of tissue, as in the case of the acids, *beyond the diseased parts*.

These constitute the most practical points in the treatment of this important affection ; and I have, as far as possible, confined myself to giving you what I have found the most efficacious remedies, without cumbering your minds with a quantity of useless prescriptions.

LECTURE III.

THE INITIAL LESION OF SYPHILIS.

To-day we break ground upon the most important venereal disease which can afflict mankind, important not only from its effects upon the original bearer of the disease, but also from the horrible consequences which may be entailed upon the offspring of the syphilitic person; and in dealing with syphilis I shall try to give you, as clearly and practically as I can, the chief points of the disease, and in what its first symptom, the initial lesion, differs from the chancroid.

In the first place, let me explain why I abandon the name chancre. First, because it is *confusing;* and second, because it means nothing. The French, English, and most American writers call the syphilitic sore, chancre, and the local venereal sore the chancroid; but the Germans expunge the word chancroid from their vocabulary, calling that lesion chancre, and our chancre the initial lesion of syphilis; and this multiplying of names is confusing. Chancre, originally derived from cancer, means "something which eats." Now, the initial lesion does not do this, and the word chancre does not necessarily mean anything syphilitic; but to say *initial lesion of syphilis* means that it is the first symptom of syphilis.

And bear this well in mind, it is syphilis already; no local lesion, as was the chancroid, but the first symptom of a disease which is always serious, sometimes grave in its results, and connected with other symptoms which do not appear for some weeks after. I shall therefore call the *first symptom of syphilis the initial lesion,* and entirely abandon the word chancre.

The first case I have to present is of interest in several ways; and before commenting at length upon it, let me give you a few points in the history : The patient, a stout, well-built young fellow, twenty-four years of age, was admitted to the hospital, November 7, 1879. He says he has had gonorrhœa and chancroid several times, but you observe syphilis is not included in the category. A very noteworthy omission. Very rarely indeed does a patient contract syphilis more than once in a lifetime ; chancroid and clap can be caught *ad libitum.* But to go on with the history : on the 9th of August, 1879, he contracted the present sore, thirteen days, he declares, after the connection. Here let us pause; *thirteen days after coitus the sore breaks out.* You remember what we found to be the case in studying the chancroid, " the sore came on two or three days after coitus ; " here it is thirteen — four to six times longer. Deduce, then, this axiom : *the initial lesion of syphilis is endowed with a period of incubation which is denied to the chancroid.* But there is something still more interesting in this thirteen days' incubation. As a rule, the incubative stage of the initial lesion is longer—on an average 21 days; but this period is variable. If, then, we reckon 21 days as the average in such cases, 13 days, the present stage of incubation. is shorter than the usual time, although not the shortest recorded. The limits which are now recognized are maximum 98 days, minimum 10; and although these represent extreme cases, bear the possibility of their occurrence in mind in making your diagnosis.

To formulate the matter in a few words, *always suspect the nature of a venereal sore which has not appeared until ten days or more after coitus.*

The history goes on to say that "it (the sore) commenced on the under surface and on the right side of the prepuce, and the soreness, swelling, and induration came on within four days. At present he has an induration extending all over his prepuce."

The induration, which is very perceptible, is under the finger of a hard, resilient character, entirely distinct and separated from the surrounding tissues, and is seated upon a non-inflammatory base. Contrast this with what we found in the chancroid. In the latter the tissues were soft and supple; there was *no induration*, and the ulcer was angry-looking — inflamed, in other words. In the initial lesion under observation, the ulcer, if indeed we can call it an ulcer, is very *superficial ;* it looks more *like an erosion ; the floor is clean and red in hue, the edges* sloping and not undermined.

Another point of interest is the fact that this variety of venereal ulcer does *not* have any tendency to *spread nor to eat* into the tissues, as does the chancroid ; indeed its whole course is *cold and slow*, and shows, nine times in ten, a *greater inclination to heal up than to extend*—another point of difference between it and the chancroid, where we found the opposite attributes.

Besides this, we observe the *singleness* of the lesion and the *scantiness of the secretion* as noteworthy conditions of difference between the two varieties of ulcer. With regard to the singleness of the lesion, you remember we found in chancroids that *multiplicity* was *not exceptional*, and that this was brought about in two ways : either *as independent foci* of infection, or by *auto-inoculation ;* but in the *initial lesion of syphilis multiple sores are the exception* rather than the rule, *and when they occur it is as independent foci of infection, never from auto-inoculation. Bear in mind, then, that the secretions of syphilis cannot be inoculated as syphilis upon a syphilitic person.*

The nature of the secretion is also deserving of a few words ; it is *thin* and *scanty, not abundant* and *purulent*, as we find it in chancroids, and unless the ulcer is irritated from some cause, *never becomes purulent.*

I wish now to. call your attention particularly to the *induration,* for this is a very important point, and one upon which too much stress cannot be laid. Whenever this symptom is found *clearly* and *well marked,* it is of *value* as stamping the lesion with a character. But there are many cases in which the induration is *very thin and slight* (parchment induration); nay more, where the *induration is entirely wanting.* Yet the sore has *not* changed its nature; it is still *syphilis, and will be followed by secondary symptoms* as certainly as is the hard variety. This is why I urged you, when speaking of the chancroid, to abandon the use of the word "soft;" for if you regard the soft sore as the one which is *par excellence* local, and does not infect the constitution, what are you going to say of the sore which does contaminate, or, to be more strict, which is the first symptom of systemic contamination, yet which is "soft"? Pray what does the name tell you? Nothing; but chancroid and initial lesion do mean something; they tell you that the first is a local disease, the second a constitutional one.

The term "hard sore" is also objectionable, because the hard sore means syphilis in contradistinction to the "soft sore," which means the opposite; and yet some soft sores are syphilis. No! I think the names I give you are the best; if you know better ones adopt them ; if not, use these with me.

Let me then give you another formula :

The initial lesion of syphilis is usually indurated; when present, this is of great value ; but its absence, which sometimes happens, does not change the nature of the lesion ; it still remains syphilis. When the induration is absent, the diagnosis has to be made from other characteristics.

We will now pass on to study the condition of the glands in the commencing stage of syphilis, and here we shall find many points of difference between the initial lesion and the chancroid.

To go back a little ; you remember in studying the chancroid we found that the inguinal glands were thickened and brawny — confounded, so to speak, with the surrounding tissues, in such a manner as to make a doughy mass, which showed, moreover, decided inflammation. Turn to the cases before us, and what do we find ? The glands in the groin are enlarged, it is true, but they are perfectly distinct from one another ; they roll about under the skin freely and easily. When handled they are not fused together nor with circumjacent tissue, as is the case with the chancroid, and they are painless.

Could anything be more opposite than these two kinds of bubo ; yet this is not all. Syphilitic buboes *rarely suppurate ;* when they do, it is from some other cause than the syphilis—generally from debility or an

2*

enfeebled constitution, and the *pus* they furnish is *laudable* and *incapable of conveying the disease either to the bearer of the lesion or to others ;* in other words, they are simple abscesses, such as you are liable to meet with in any person who is run down in health. Neither are they dependent upon the site of the initial lesion, but are met with on both sides of the body, and are due to the systemic poisoning which has occurred—to the same cause which has produced the initial lesion itself, and *not to absorption of matter from the ulcer*.

When I come to speak of the subsequent syphilitic symptoms, I shall show you how the glands over the body are similarly enlarged—what is called the *adenitis universalis syphilitica*.

Of the initial lesion of syphilis, there are several varieties; the archetype, sometimes called the Hunterian induration, you have already seen. You can tell it as far as you can see it, and it is unmistakable, but unfortunately it is not always present. Sometimes the *initial lesion* has but a thin disk-like layer of induration beneath it, which gives to the finger the sensation of a slight layer of parchment beneath the skin or mucous membrane—the " parchment induration " which I have already brought to your notice ; and again, very rarely it is true, there may be no induration at all. The ulceration in the initial lesion is usually very superficial, and, when seated upon a markedly indurated base, is raised above the surrounding tissue ; it is then known as the *ulcus elevatum*, and again it may be a mere erosion which, conjoined with little or no induration, is very puzzling, and apt to mislead the surgeon as to its true character. Beware of such ; do not be in a hurry to pronounce positively on the nature of any such lesion, but suspend judgment, else you may make an awkward mistake by calling a given lesion innocent, which a few weeks later will be followed by a general outbreak upon the skin and mucous membranes. In addition, the initial lesion has no destructive tendency, no undermined edges, no gray floor ; on the contrary, it has a red granulating appearance, with oftentimes a dark spot in the centre, and is prone to bleed readily upon handling.

In those cases where the initial lesion itself gives little or no information, appeal to the chain of glands nearest to the lesion. You will seldom find them intact, and their induration will often help you to a diagnosis.

Let me, before going further, make in tabular form a comparison between the initial lesion and the chancroid :

CHANCROID.	INITIAL LESION.
Little if any period of incubation.	Decided period of incubation.
Destructive, with tendency to spread.	Not destructive ; tends to heal rapidly.
Edges undermined.	Edges sloping, not undermined.

CHANCROID.	INITIAL LESION.
Copious, purulent secretion.	Scanty, serous secretion.
Contagious and auto-inoculable character of the pus.	Secretion not auto-inoculable.
Usually multiple.	Usually single.
Not seated upon an indurated base.	Generally indurated; sometimes—rarely, however—not.
Glands liable to become inflamed; when so, they may suppurate and become a chancroid, furnishing inoculable pus.	Glands indurated, not inflamed; very rarely suppurate, and then from causes apart from syphilis. Never furnish inoculable pus.

This gives you, at a glance, the important points of difference between the two ulcers.

The site of the initial lesion is a point of much interest, and I wish to recall to your minds what I said in an earlier lecture about some forms of venereal diseases being transmitted without sexual contact. This is the case in syphilis—the initial lesion not infrequently being met with upon the lips, the cheek, or upon the nipple: in the first two cases from kissing or from using contaminated utensils, a pipe, a spoon, or drinking vessels; and in the latter from suckling a syphilitic child. Other places are the fingers, the nose, the tongue, the throat, and the palpebral conjunctiva of the eye; in short, lay it down as an axiom, *that no portion of the body is exempt from being the seat of the initial lesion, although the genitals are the usual seat, and naturally so from being more exposed.*

The source of infection is another point to which

I invite your attention. A chancroid, as I have already explained to you, comes from a chancroid or a chancroidal bubo ; but syphilis is caused in other ways than from inoculation of the secretion of an initial lesion. The secretion from mucous patches, whether of skin or mucous membranes, as well as the blood of a syphilitic, during the first twelve months at least of the disease, is capable of infecting a sound person, but, as I have already told you, it is not auto-inoculable. The tears, saliva, and sweat are innocuous, and until within a few years human milk was included in this category, but some recent experiments have made this doubtful, although the reported cases are by no means convincing. It is the contagious property of blood and mucous patches which cause many of the cases of initial lesion of the lips, cheeks, and nipple ; the patient, not being aware of the danger, kisses healthy persons who perhaps have an abrasion of the lips, and the disease is lighted up in them. As regards the nipples, the mucous patches of the baby's mouth perform the same office for the nurse.

Suppose the infection to be derived in one case from the secretion of an initial lesion, in a second from that of a mucous patch, and in a third from syphilitic blood ; how does the disease begin in these cases ? *Always by an initial lesion seated at the point where the virus gained entrance, never in any other way.* Syphilis does not first make its appearance in the form of a so-called secondary eruption without a preceding initial lesion, although there are some cases where this would seem to be so. These cases are when the initial lesion is seated in some unusual or not readily accessible place—as, for example, in the urethra of the male, in the cervix uteri, upon the lips or fingers of both sexes. When it is seated in the urethra, palpation often reveals the remaining induration, and sometimes separation of the labia urethræ reveals the syphilitic erosion; and a slight, gleet-like discharge is also present.

Another cause of confusion, when the patient has not come under observation until after the outbreak of general symptoms, is that the initial lesion becomes changed into a mucous patch—a symptom of the so-called secondary stage; but even here the traces of the induration will put you upon your guard as to the real nature of this supposed mucous patch.

The initial lesion is also subject to complications, though to a less extent than the chancroid; the principal ones being phimosis and phagedena. When phimosis attacks the initial lesion it is not so likely to produce such serious consequences as when it occurs with the chancroid, owing to the inflammation being much less, and also to the fact that the initial lesion does not ulcerate. The only danger to be apprehended from this complication is gangrene, and that may be so readily and easily obviated by an incision as to practically rob it of one-half its danger. You note that I said '' easily obviated by an incision,'' and I wish you here to remember what was said in regard to this complication when speaking of the chancroid. Then I advised you not to cut, unless obliged to, because the edges of the wound would become chancroidal; but in the initial lesion no such danger is to be apprehended; the secretion of the lesion and the blood of the syphilitic are incapable of being auto-inoculated. You may therefore operate, if you see fit, at once, so far as contagion is concerned, but I should advise waiting a little, for the following reasons: first, because no operation should be done if the same result can be attained in any other way; and, secondly, because the induration, even if very thick and marked, will disappear under proper treatment and with it the phimosis. But should gangrene threaten, then you not only may, but should operate to avert this threatened evil, and you may practise the single or the double incision already advised in Lecture II.

Phagedena, in syphilis, is of as grave import as in chancroid, and comes from the same cause, viz.: constitutional defects, due to alcoholic abuse, or to a morbid diathesis, and it plays an important part as regards prognosis. The ulceration, instead of being superficial, then becomes deep and wide-spread, the floor is gray and pultaceous, the secretion more abundant, and the induration may entirely melt away under the phagedenic action. Where the initial lesion is phagedenic the subsequent lesions are apt to take on ulceration, and to pursue a rapid course, being rebellious to treatment, and exposing the patient to grave and serious consequences.

Before going on to speak of treatment, let me say a few words about what is generally called the "mixed sore." I wish the term could be abandoned, as it is confusing and does not convey a correct idea of the facts. It is really a double sore ; there is no mixture whatever of nature, course, or virus ; it is simply where inoculation of a chancroid and syphilis occur simultaneously in the same person. The two poisons being received at the same coitus, they operate differently as regards the time of their appearance. The chancroid appears first ; remember, it has no period of incubation, and runs its course and perhaps gets well before the initial lesion comes upon the stage. At a later period, usually varying from ten to twenty-one days after the infecting coitus, the initial lesion appears, marked by its peculiar characteristics. It sometimes happens that the chancroid has not healed before the first symptom of syphilis is due. This, then, is what happens : the chancroid is surrounded with a ring of induration, the secretion becomes less copious, the floor fills up and appears redder and healthier, and the nearest chain of glands is indurated ; the chancroid has, in other words, become changed into an initial lesion. But through the whole performance there is no interchange of characteristics, the two lesions remain entirely distinct, and " mixed chancre " is, to my mind, a misnomer ; I prefer to call it a double infection, double in the sense that two kinds of virus have been deposited in the same spot.

It is in these cases of double infection that you will be most likely to meet with a suppurating bubo, the pus of which is auto-inoculable, and which, unless you are forewarned, may lead you to believe that syphilis is attended with a suppurating, auto-inoculable bubo. The bubo is chancroidal, similar to what we have already studied, has nothing to do with the syphilis, although it is contemporaneous with the initial lesion, and will require the treatment appropriate to chancroidal buboes.

As regards treatment it is simple and, so far as the local trouble is concerned, effective in the majority of cases. In the first place, let me beg of you *never to cauterize an initial lesion unless it should be attacked by phagedena.* I know it is the rule to cauterize every suspicious-looking ulcer, but in the case of the initial lesion it not only does harm in irritating an otherwise simple ulceration, but it retards its healing. Dress the lesion simply; sometimes a piece of lint laid over the ulceration or erosion will suffice, but at other times a little more active treatment may be requisite. Of all dressings I much prefer the dry, and of them iodoform heads the list, either alone or in combination with other drugs. Thus—

℞. Iodoform pulv.
Lycopodii pulv. p. æ.

Or—

R. Pulv. zinc. ox. ....2 parts.
Pulv. iodoformi.................1 part.

Or—

R. Pulv. hydrarg. chlor. mit...........1 part.
Pulv. iodoformi.................2 parts.

Calomel, without anything else, may also be used with advantage.

A mode much practised in the German hospitals is to apply a piece of the Emplastrum de Vigo cum mercurio, the size of the ulcer, directly upon the sore, and leave it thus protected from the air, until the ulcer heals up. The Emplastrum hydrargyri, U.S.P., will answer as well.

If you prefer to use a wet dressing, a weak solution of carbolic acid is the best, of which the following will serve as an example:

R. Cryst. ac. carbol.................gr.ij.
Aquæ..........................℥ iv.
M.

Apply on lint or cotton thrice daily.

Constitutional treatment, whether internal or external, is better not employed, save in exceptional cases, until the subsequent (secondary) symptoms appear, because, in many instances, it is impossible to diagnosticate the nature of the lesion under observation, and inasmuch as mercury, when given during the existence of the initial lesion, has a tendency to retard the outbreak of the secondary symptoms, it leaves the surgeon in doubt as to what the disease really is, and unable to tell his patient what or what not to expect. Delaying until secondary lesions come on, or until the period at which they should appear has passed, does not injure the patient's prospects of recovery, and it does give the surgeon the opportunity of informing his patient as to the nature of his disease.

There are cases where it is necessary to cure the initial lesion rapidly, as for instance in married people, and to retard, and as far as possible check the subsequent manifestations; but in such cases the patient should be told that by so doing the surgeon will be unable to tell him or her what subsequent symptoms to expect, or to count upon probable recovery, even after many months of treatment.

These exceptions do not then conflict with this general law, viz., *do not treat the initial lesion by the internal use of mercury, but await the development of secondary symptoms.*

Internal treatment by tonics, iron, quinine, and the like are admissible in this stage, should the patient be anæmic, a very frequent condition in syphilis.

LECTURE IV.

IN the last lecture we passed in review the initial lesion of syphilis, dwelling upon its characteristics and the main points of difference which exist between it and the chancroid. This lecture I propose to devote to considering the nature of the subsequent lesions which occur in syphilis, what are commonly known as the secondary and tertiary symptoms, more particularly those which occur upon the skin, reserving the syphilides of the mucous membranes to a subsequent occasion.

In the first place as regards the nomenclature: I wish you to remember that the terms secondary and tertiary are ones of mere convenience, and must not be accepted in a purely chronological sense. Many of the symptoms which are classed as tertiary, may, and do, appear in the secondary period, as, for example, the affections of the nervous system, and, should you be too bound down to name and rank all affections of the nervous system as necessarily tertiary, you will involve yourselves in much confusion and trouble. The true distinction I believe to be this, viz. : that during the secondary stage the symptoms are more superficial, and more amenable to treatment than they are during the tertiary period, and that the exudations which occur during the earlier stage, are absorbed and removed more speedily than those of the latter. In addition to this they have not the same destructive tendency, for we shall find as we go on that the tertiary lesions are marked by deep, and oftentimes serious, loss of tissue, while the secondary lesions are, comparatively speaking, mild, and leave behind no traces of their presence. I therefore much prefer to speak of these lesions as the superficial and the deep lesions of syphilis, irrespective of their seat, whether on skin or mucous membrane, in the eye, ear, nervous system, or bone.

Before the symptoms upon the skin and mucous membranes appear there is a period of rest (incubation) between the occurrence of the initial lesion and the advent of the subsequent manifestations, during which time the initial lesion may have entirely healed up, leaving only the induration of its former site, and an induration of the nearest chain of glands as traces of its presence. Even these latter may be very indistinct, rendering the connection between the two sets of symptoms vague and uncertain, and the relation each bears to the other would be overlooked unless you were forewarned. Note, then, that there are *two periods of incubation in the early stages of syphi-*

lis, the first being between the infecting coitus and the appearance of the initial lesion, and the second between the appearance of the initial lesion and the coming on of the early syphilides.

The length of this incubative stage varies within certain limits, as does the incubation of the initial lesion. For all practical purposes you may consider the maximum limit as about ninety days, or three months, the average being forty-two to forty-five days, or between six and seven weeks. The minimum limit you may fix at twenty-five days, or between three and four weeks, just about the length of the incubative period of the initial lesion.

Formulate for yourselves, then, this rule : *the early syphilides have, like the initial lesion, a period of incubation, the average length of which is forty-five days, but which may extend to ninety, beyond which time it is rarely protracted, unless it has been prolonged by the internal treatment of the initial lesion with mercury.*

Before the early syphilides make their appearance there are certain vague and by no means constant symptoms which precede them by a few days, and which are known under the name of " prodromata." These are *fever, rheumatoid pains* of the muscles, *aching of the bones*, especially of the superficial long bones, such as the ulna and the tibia, and *headache*, usually confined to one lateral half of the head (hemi-

crania). The peculiar feature of these symptoms is
that they come on *at night* when the patient is in bed,
but not until the heat of the body has warmed
the bed; hence in those patients whose occupations
oblige them to turn day into night, such as bakers,
the pains come on in the daytime, when they are
warm in bed—so it seems to be the heat really
which brings out the pains and not necessarily the
time. When the patients are up and about, these
symptoms vanish. During the fever there may also
be a slight rise in temperature, although this is not
constant.

After the prodromata have lasted for a few days,
the syphilides make their appearance upon the skin
and mucous membranes, and the first of these is the
erythema syphiliticum, or, as it is commonly called,
"roseola." And here again I wish to protest against
the names which have been commonly given to these
syphilides of the skin. They are, roseola for the
erythemata, psoriasis for the scaly eruptions, which
come upon the palms of the hand and soles of the
feet, ecthyma and rupia for the pustulo-crustaceous
manifestations of the later stages of syphilis, and are
borrowed from the slight resemblance they have to
the corresponding non-venereal eruptions which appear on the skin. The objection I make to these
names is that they are complicated and confusing,
and I much prefer the nomenclature I shall presently
give you as being simpler, and more accurately de-

scribing their pathological condition. The names I propose for your use are—

Erythematous,
Papular,
Papulo-squamous,
Pustular,
Pustulo-crustaceous.
Tuberculo-crustaceous, ⎫
⎬ Syphilides.
⎭

Ulcerating
and
Non-ulcerating ⎬ Gummata.

This includes all the varieties of the syphilitic man-ifestations of the skin, and the advantage of their names is that it describes accurately the pathological condition of the lesion and the cause at the same time. Thus a papulo-squamous syphilide, although a little longer than syphilitic psoriasis, tells you more, and the same is true of pustulo-crustaceous syphilide as against syphilitic ecthyma. I shall, therefore, in de-scribing the syphilides of the skin, use the above nomenclature, and the one which heads the list is the erythematous syphilide. .

VARIETIES : ERYTHEMA MACULATUM—ERYTHEMA PAPULATUM.

Erythema maculatum.

This is the first one of the skin eruptions to make its appearance, coming on about forty-five days after the initial lesion, and is characterized by rose-colored blotches, not elevated above the surrounding skin, abundant over the entire trunk, arms, and legs, sometimes invading the face, notably the forehead, and occasionally being met with on the inside of the hands and on the soles of the feet. Just before the rash fully declares itself, there is a peculiar mottling of the skin, looking as though the eruption were under the cuticle but had not yet made its way through. There may be at this time some *nocturnal syphilitic fever*, with a slight increase of temperature. One other symptom I have reserved, as I wish to dilate a little at length upon it, and that is, there is *no itching*. Syphilitic eruptions do *not* itch, although the skin of syphilitics is often irritable, hence, if you inquire of such if there be any itching, they will as likely as not reply in the affirmative, yet when you come to examine the skin there are no marks of finger nails, as are found in phtheiriasis, eczema, lichen, etc. Do not, therefore, be thrown off your guard by any supposed itching of the skin in syphilis (of course if lice are present the case is different, but their presence and subsequent removal will explain and cure this symptom), although there may be. especially in women, an irritability of the derma.

The erythematous syphilide pursues its course evenly and quietly, passing on from the distinct rose-colored stains to a coppery hue, then to a dingy yellow, and finally disappears entirely with a slight des-

quamation of the cuticle, leaving no trace of its presence. A few words about the *coppery hue* of the
syphilides. Its diagnostic importance has been
much exaggerated, and you will, in many non-
venereal skin eruptions, see as much of the copper
color as you will in the syphilides.

Erythema papulatum.

This variety of erythema comes on after the macular kind, sometimes even before its entire disappearance, and seems to be the intermediate link between
the erythemata and the papulæ. It is raised above
the level of the skin, is flattened and seated upon a
broad base, is of a darker hue than its congener, the
erythema maculatum, and always more or less scaly.
This desquamation in the syphilides is somewhat
different from what takes place in the simple kinds
of eruptions; it is *rather a peeling than an actual
scaling.* It is less widely distributed than the macular kind, being found chiefly on the back of the neck,
on the back, and on the volar surfaces of the arms
and legs; it also affects the palms of the hands and
the soles of the feet rather more than the E. maculatum. Not infrequently, as I have already said, it is
found conjoined upon the body with the macular
variety; indeed, in certain parts of the body where
heat and moisture are found, the macular seems
rapidly to pass into the papular eruption. This is
especially noticeable about the genitals of women

where these papules become quite luxuriant in their growth and secrete abundantly ; perhaps you recall several cases of the kind which I have already shown you in the wards. This variety paves the way to the next stage in the disease where papules take the place of the erythemata.

VARIETIES : PAPULÆ MILIARES—PAPULÆ LENTICU-
LARES.

Papulæ miliares.

The course which the erythemata pursues varies somewhat according to the intensity and the amount of acuteness of the disease. Sometimes the erythema will entirely disappear, leaving the skin unblemished, and this freedom from disease may last for some weeks before the next step is reached. Here you see then a tendency to incubation even between the various kinds of the eruption, but sometimes the attack is much more rapid than this, and before one form of eruption has gone another comes on, so that upon the same subject you will find macules, papules, and even pustules scattered over the body, constituting what is known as "*polymorphism.*" Remember then that the papules may not appear until several weeks after the disappearance of the erythemata, especially if a mercurial treatment has been instituted, or it may come " with a rush,"

so to speak, one train of symptoms crowding upon
the other, leaving no interval of repose or apparent
freedom from the disease.   We will suppose that the
macules have disappeared and that the papular stage
is due; what must we look for ?   The *nocturnal pains*
which had become almost nil now *return*, and there
may be *some fever*, when suddenly over the entire
body, arms, legs, face, and scalp, small pointed ele-
vations of a reddish color break out, which are closely
packed together, and are sometimes crowned at their
apices with a small scale.   These go on for several
weeks getting more and more purple in hue, the
papules become more scaly, less elevated, and
finally melt away, leaving a staining, which from
being at first purple becomes a yellowish brown,
and this in its turn is absorbed, leaving the skin free
from scar or blemish of any kind.   These papules are
small and bear some resemblance to the simple acne
which invades the face and shoulders, save that they
are much more numerous ; hence the name sometimes given it, of acne syphilitica.   One peculiarity
about this eruption—indeed, you may say about all
the syphilitic eruptions—is their tendency to assume
a circular form, grouping themselves in the shape of
a ring or segments of a ring over the body, and this
is kept up even into the late stages of the disease.

When these papules are seated upon the forehead,
they assume somewhat the appearance of a ribbon or
band stretched from temple to temple, and among

the older syphilographers received the fanciful name of "Corona Veneris," a by no means inapt title. These papules extend into the hairy scalp, where from irritation of the comb and finger-nails the apices become covered with a bloody scab, somewhat resembling the disease of the scalp called impetigo capitis. We shall see the same thing occur in the pustular stage of syphilis, except that there the crust is larger and thicker.

Papulæ lenticulares.

After the miliary papules have run their course, sometimes even before, the next variety of the same eruption, the lenticular, manifests itself in the shape of broad flat papules, considerably raised above the surface of the skin, of a color similar to the preceding, but covered with a thicker and darker scale, which occasionally become transformed into a very thin crust, due to exudation from the papule • itself. These papules are not widely disseminated over the body as are the erythemata or the papulæ miliares, but are found in isolated groups upon the palms of the hands and soles of the feet, between the fingers and toes, upon the genitals of both sexes, at the angles of the mouth, where they are frequently continuous with mucous patches of the buccal cavity, at the edge of the hairy scalp, upon the shoulder blades, on the buttocks and thighs. When grouped together, as they often are, and covered with scales they bear some resemblance to patches of psoriasis vulgaris, but in this latter disease the scales are of a more silvery white color, and are smaller than is the case in syphilis. When found upon the genitals and between the toes, the heat and moisture of the parts favor their growth and development; they lose their scales, and the secretion which exudes from them covers their surfaces with a dirty white layer, which can be wiped off, revealing a glazed red floor. These are the lesions which have been described as *mucous patches of the skin*, but which the Germans more accurately call the " moist secreting papule."

When seated at the junction of mucous and cutaneous surfaces on the genitals the papule retains very much the same characteristics as the mucous patch, but at the angles of the mouth the skin lesion from exposure is covered with a dry scale, sometimes a thin crust, while the lesion of the mucous membrane is moist and covered with a whitish pellicle.

But it is to their position upon the palms and soles that I wish to invite your special attention. In the beginning of their growth the papules are broad, flattened, and of a deep purple color, the apices are covered with scales, which are renewed as soon as they get rubbed off. Later on these papules coalesce and form broad patches, which become fissured and bleed, and the blood mingled with scales forms a thin crust upon the surface of the lesion. These patches extend in size, become very much thickened, and covered as they are with scales and dried blood, are often, with difficulty, distinguishable from chronic eczema of the palms of the hands. But I beg you to bear in mind that this latter affection is, in my experience, a rather uncommon disease, whereas a papular syphilide of the palms is not infrequent. Of course, when you are able to get a history of syphilis, the nature of the lesion is clear, but sometimes you may get none, perhaps cannot ask for any, and in such cases it will stand you in good stead to remember that nine times in ten such lesions of the hands and feet mean syphilis.

Formulate, then, this rule : *Squamous affections of the palms of the hands and of the soles of the feet are nearly always syphilis and require anti-syphilitic treatment.*

We are now nearing the boundary line which is supposed to separate the secondary and tertiary lesions, and heretofore we have noticed no tendency to ulcerative destruction ; all the lesions go off and leave no trace behind them. But in the next stage this is changed, pus is formed, and pus means destruction of tissue. The lesions which we are now to consider I have divided into two groups, *the pustular* and the *pustulo-crustaceous, i. e.*, those which remain pustular, not becoming covered with a crust, but being absorbed, and those which break down and are covered with a scab.

PUSTULÆ.

This variety begins differently from any which we have heretofore examined, having its seat more deeply embedded in the tissues than the papule, and starts from the true skin and not in the epidermis. It is the kind known in the books as "impetigo syphilitica." Starting then from the deeper layers of the skin, it is felt beneath the surface as a small hard point, which rapidly becomes elevated and is crowned at its apex with a pustule. This pustule increases in size, and may occupy the entire base upon which it is seated, said base being surrounded by a purple areola, while the pustule itself is yellow. This pustule is full, round, and in the majority of cases distended with matter, which, if the pustule is broken, dries into a small superficial crust, revealing on removal a slight ulceration beneath. Moreover, this pustule is not umbilicated, as is the case in variola. Provided the course of the disease is favorable, the pustule dries up and becomes covered with a few flakes of dried epidermis ; these are subsequently cast off, and a discoloration of the skin remains. After a longer or shorter time this staining fades away, and unless the pustule has started from deep down in the tissues no scar is left behind. If its origin has been deep-seated, after the pigmentation vanishes, a white scar is visible, corresponding to the size of the pustule, and is due to an atrophy of the cellular tissue beneath the skin. This is not so marked as it is in the crustaceous syphilides.

These pustules are widely scattered over the body, the head, face, trunk, arms, and legs being invaded, resembling, in this respect, the erythematous and papular syphilides. This variety may be succeeded by another crop of pustules of the kind I have here designated as crustaceous, and which are more serious than the ones we have just studied, inasmuch as they are always attended by ulceration, sometimes quite extensive, and are not so amenable to treatment.

Pustulæ crustaceæ.

The pustulo-crustaceous syphilides commence with a more pronounced and more diffused amount of exudation beneath the skin than do the non-ulcerating pustules ; they come rapidly to the surface, the pustule breaks, and when it does an ulceration more or less extensive is found beneath. Sometimes this ulceration does not penetrate deeply into the tissues, but spreads laterally over quite an extent of surface, secretes abundantly, and presents irregularly shaped borders (scalloped), due to the coalition of several individual pustules or groups of pustules. This is known in the books as syphilitic ecthema. At other times the pustule increases enormously in size, ulcerates, and the ulceration extends deeply into the tissues, making a punched-out cavity, which is covered

3*

over by a thick brown or black crust, due to the admixture of blood with the pus. This crust continually increases in height from accretion, at its base,
of fresh matter from destruction of tissue, and forms
over the ulcer a conical scab, from one-half to two
inches in height, which is firmly mortised into what
seems to be sound skin, but which, on removal of the
crust, is seen to 'be undermined by the ulceration.
This undermining of tissue is also found in the so-
called ecthymatous variety, but in a much less degree. This is the rupia of the books, one of the
worst forms of the syphilides you will be called
upon to deal with, and which is frequently rebellious
to treatment.

The seat of both varieties is more limited than is
that of the non-ulcerating pustules, and when they
appear are more likely to be discrete. The face, the
upper arm, the thighs, and the buttocks are their favorite situations, although they are sometimes found
upon the trunk, especially the back.

Closely conjoined with the pustulo-crustaceous
syphilides in nature and course are the tuberculo-
crustaceous eruptions. They affect the same portions of the body as the former, and only differ in
their commencement in being larger and harder, and
may be regarded as the connecting link between the
pustule and the gumma. The ulceration which ensues runs much the same course as in the so-called
rupia—is deep, destructive, and often rapid, the crust

is thick and elevated, and in its subsequent course is not to be distinguished from its congeners of the pustular variety. When I come to speak to you about the syphilitic affections of mucous membranes I shall show how, under certain conditions, these ulcerating syphilides may be mistaken for chancroids.

The next and last symptom to be spoken of is the gumma (pl. gummata), in which the amount of infiltration into the skin and cellular tissues is very abundant and brawny, and if it breaks down gives rise to a very serious and nasty-looking ulceration. Two varieties of this gummous infiltration exist, the diffuse and the circumscribed, and both kinds, if left untreated, will ulcerate. The resulting sore is deep, has a tendency to burrow, has a yellowish floor covered with the remains of dead and dying tissue, and secretes abundantly—in many of these points resembling a chancroid, but in their nature they are entirely dissimilar. A chancroid becomes worse under a mercurial course ; it is poison to it, while, in the lesion under consideration, mercury is the only thing that will do it permanent good.

These gummata are found upon the thighs and arms more frequently than they are elsewhere, and are single rather than multiple, although they may be associated with gummata in the viscera and in mucous membranes. When patients have arrived at this stage of visceral syphilis a very peculiar condition of the system supervenes : what is known under the name of "syphilitic cachexia." In this stage they steadily but surely run down, the functions are no longer active, assimilation either of food or medicine ceases, and death supervenes from exhaustion. Happily such cases are rare, but their occurrence serves to show what syphilis is capable of doing.

We have now finished the study of the lesions known as syphilides of the skin, and I have given you the salient points found in them without burdening your minds with unnecessary details. I have passed over two varieties in silence : the vesicular and the bulbous syphilides, which are described in some treatises on venereal diseases. I omit them for two reasons—first, because I doubt their separate existence (both of them really belong to the pustular syphilides); and secondly, if they do exist they are so *very* rare as to make them curiosities of syphilis rather than regular lesions, and my object in these lectures is to avoid undetermined points and to give you only what is practical and certain. But before passing on to a consideration of the effects of syphilis upon the appendages of the skin, I wish to say a few words as to the general course which the cutaneous syphilides pursue.

In the first place, after the initial lesion has passed away, there may be a period of apparent immunity from the disease before the syphilides appear ; this I have already told you is the period of incubation between the so-called primary and secondary stages.

The erythemata appear and disappear, leaving another intermission between the erythemata and the
papulæ, and this period varies from two weeks to one
or two months, according to the activity and efficacy
of treatment. After the subsidence of the papules,
another period of repose of several weeks may occur
before anything further appears, when some variety
of the papular syphilides will recur, or, if the disease
is progressing, pustules will show themselves. So it
goes on, each stage advancing progressively from
superficial to deep lesions — from those symptoms
which are mild and which are readily absorbed, to
those which are ulcerative, destructive, and which
are not absorbed.

But, in place of advancing, we will suppose the
disease yields to treatment; what do we see then?
The erythema vanishes, and the patient, though kept
under observation for some time, displays nothing
more ; or, at the end of several months, he may show
a slight recurrence of the erythema, or perhaps a few
scattered papules. Treatment is vigorously pushed,
the papules disappear, and the patient hears nothing
more from his syphilis. He is, to all intents and purposes, well. But there is one point I wish to lay
stress upon : *syphilis never runs a hap-hazard course*,
it never begins with deep-seated lesions first, to show
later on superficial ones, but it pursues, if a serious
case, a pretty steady course from bad to worse ; if,
on the contrary, it be a light case, occasional relapses

of the same kind of eruption may occur, but it never skips about. I shall revert to this point again when I come to speak of the prognosis.

As regards the course these lesions pursue, you may lay down this broad general principle: the superficial lesions disappear pretty quickly, the deep-seated ones quite slowly. In order that you may comprehend this readily, I append here a table giving approximatively the time after the appearance of the initial lesion at which the various syphilides are due, and their duration.

NAME.	DUE.	DURATION.
Erythema....	6–12 weeks............	3–6 weeks.
Papules......	2–6 months............	4–8 weeks.
Pustules.....	6–15 months...........	2–4 months and more.
Gummata....	1–5 years and more.....	½–2 years and more.

As appendages of the skin, the hair and the nails invite our attention, and of the former there are two varieties of syphilitic disease known as alopecia, one of which occurs in the early, and the other in the late stage. The early alopecia is the more general of the two, not being confined to the hairy scalp, its usual seat, but attacking the hair of the face, and even of the entire body. I have seen one case where the patient lost all the hair of his head, face, and body. This seems to be due to changes going on in the hair bulbs themselves, and not to any changes in the follicles, so that the hair grows again as luxuriantly as before. This is not the case in the late stage, when the lost hair is not generally replaced, and this is due to disease of the follicles themselves, as well as to their destruction from deep ulcerations of the scalp, face, etc.

The early alopecia is coincident with the erythema and papules, the late with the pustular and tuberculo-crustaceous eruptions.

The affections of the nails belong to the late stage of syphilis, and are usually concomitant with the pustular lesions. During the existence of the papulo-squamous syphilides, however, the nails of fingers and toes are sometimes affected; they crack, the edges become ragged and uneven, and at times scaling of the surfaces takes place. But later on in the disease, pustules occur in the matrix of the nail, causing detachment, and the nail drops off. After this happens the ulceration of the matrix may continue, destroying it and with it all hope of a renewal of the nail. If the ulceration is checked before this stage is reached, the nail may be reproduced; but its growth is very slow, the new nail is brittle, uneven, and ragged, and is seldom of much use.

LECTURE V.

FOLLOWING naturally upon the syphilides of the skin
come the syphilides of mucous membranes, and these
are among the most common of all the affections of
the earlier stages of the disease, as well as the most
obstinate to treat. They recur again and again, often being the only symptom of syphilis which remains
after the first outbreak has passed away, and are frequently a source of more annoyance to the patient
than any of the manifestations upon the skin, except
it be those of the face.

Like the syphilides of the cuticle, the syphilides of
the mucous membranes are divisible into the superficial and deep kinds, the former of which are not in
themselves serious; the latter of extreme importance,
from the consequences which they entail from destruction of tissue.

Coincident with the outbreak of the erythema maculatum, the patient will complain of a feeling of
soreness of the throat and dryness of the fauces. Inspection reveals the entire mucous membrane of a

congested red color, or, as occasionally happens, having spaces of sound mucous membrane between the congested spots, and resembling, in many respects, the eruption upon the skin.

Sometimes this erythema is continuous upon the mucous membrane of the tongue and the entire buccal cavity, and so general is it that it may be mistaken for a scarlatinal sore throat, particularly if the syphilitic fever has been at all sharp. But a little attention to the other symptoms will save the physician from such a mistake, and the treatment will definitely settle the doubt. The sides of the tongue are dotted with small punctate spots, giving it somewhat the look of a ripe raspberry, and has quite a peculiar appearance. With all this congestion there are very few physical symptoms; the voice is not materially changed, the breathing is not impeded, nor is deglutition difficult. The tonsils are sometimes enlarged, and can be felt externally as well as seen internally, and the glands of the posterior and anterior cervical regions are indurated aud slightly enlarged. In addition to these sets of glands, the following may also be implicated ; the anterior and posterior auricular, the submental and the submaxillary.

This erythema of the mucous membranes disappears in the same time and manner as the erythema of the skin, only as the parts are protected from the air no desquamation occurs. The congestion tones down from purple to red, the red to the normal pink hue of mucous membranes, and on vestige of the disease is left.

Here, also, as with the syphilides of the skin, we may have a period of rest and freedom from symptoms, but of all the manifestations of the earlier stage of syphilis, this is the most persistent, and the patient will hardly get rid of one crop of eruptions before another crop is ushered in, and that, too, while treatment is going on. Sometimes this may be a relapse of the erythema faucium, or it may be a form which I am now about to describe.

The patient consults the surgeon for a soreness of the throat resulting, as is frequently stated, from cold conjoined with " fever-sores " upon the tongue and mucous portions of the lips and cheeks. An examination shows the mucous membrane of these parts slightly thickened, as though from infiltration of the parts, and on the surface are seated opaline, glistening patches of a white color, devoid of any true ulceration, and usually sensitive to the action of hot and cold drinks, pungent condiments, etc. This tenderness is specially noticeable when the lesions are seated upon the tongue or lips. Associated with these mucous patches there may be found upon the body a papular or papulo-pustular eruption, but very often there is nothing at all except the lesions of the mucous membranes upon which to found a diagnosis. I know of few points in syphilis more puzzling to decide upon than these same mucous patches, particularly when patients insist upon their being associated with a disordered condition of the stomach, when for want of certainty as regards history and antecedents the surgeon falls into the error of considering them as simple "aphthæ."

The white covering of the mucous patches is closely adherent to the tissues below, and it cannot be detached without causing some slight hemorrhage ; indeed, in some cases, this white film is really below the surface, and is an actual infiltration into the submucous tissues with external ulceration.

This form of mucous patch is extremely obstinate, and recurs repeatedly upon the same spot, or upon adjacent parts of the membrane. Gradually, however, under active and persistent treatment they disappear, it may be for good, or else they reappear in another form corresponding to a more advanced stage of the disease.

This variety is specially to be found in the throat, its favorite habitat being the tonsils and the posterior arches of the palate. Occasionally it is found upon the dorsum and sides of the tongue, less frequently upon the buccal mucous tissue. Its first appearance is a slight elevation of the membrane from infiltration into the submucous tissue, but this does not last ; the elevation breaks down and is converted into an ulceration varying in depth according to the infiltration.

The floor is uneven and gray in appearance and

the secretion is not very abundant. But little in-convenience results to the patient from their pres-ence, as the parts become callous from the infiltration and thickening of the tissues, and the ulcers are not sensitive to heat and cold as they were in the earlier stage. These ulcers have a tendency to extend slow-ly, it is true, but still deeply, and when they are seated upon the tonsils or behind the posterior arches of the palate, they become of quite large size. It is at this stage that a change in the character of the voice takes place, and the usual clear tone is ex-changed for a hoarse whisper or an uneven strident sound. An examination by the laryngoscope shows ulceration of the mucous membrane of the larynx and of the false vocal cords with œdema. On attempted phonation it is seen the true cords do not come even-ly together, hence the calibre of the voice is materi-ally altered.

Succeeding this stage, sometimes merging into it, is the true ulcerative syphilide of mucous membranes, due to the breaking down of the gumma, which forms in the submucous cellular tissue. The first thing to attract attention is a diffuse brawny swelling of the soft parts, which progresses rapidly, breaks down, and when it occurs in those portions of the body that act as septa between cavities, it produces important and irremediable destruction. The action is rapid in these cases, a few days being oftentimes sufficient to produce extensive disfigurement. I shall return to this topic when I come to speak upon the syphilis of special organs.

In the last lecture I spoke to you of cases in which ulcerating gummata of mucous membranes might be mistaken for chancroids. A patient who has been the subject of an old and long-standing syphilis will present himself to the surgeon with a circumscribed hard tubercle seated upon the mucous membrane of the penis, either in the fossa glandis, on the reflex layer of the prepuce, or at the junction of the frænum with the fossa. This tubercle is perfectly painless, unattended with any inflammation, and apparently indolent in character. It will suddenly break down, become converted into a deep, punched-out ulcer, corresponding in extent with the original tubercle, presenting a yellow, uneven floor, devoid of induration, and secreting a thin, viscid fluid, which, from irritation, will become purulent. If this lesion be seen for the first time in the ulcerated stage, it may readily be mistaken for a chancroid, especially as it evinces destructive tendencies, for it may eat away the frænum, burrow into the urethra, and extend far beyond the limits of the gumma which gave it birth. These are puzzling cases to decide upon ; the history will sometimes help you to a diagnosis, but of all things the treatment will be the experimentum crucis.

The ordinary remedies for chancroid are useless ; cautery and local dressings do not produce the results they should, and you begin to despair. Change

your tactics, and without giving up topical applications, except the cautery, put your patient upon a
mixed treatment (mercury conjoined with the iodide
of potassium), and the result will, I know, gratify you ;
the lesion will get well.

Conjoined with these symptoms of the skin and mucous membranes, during the earlier stages of syphilis,
are others fully as important for you to know about
and remember. I refer to the enlargement of the
glands over the entire body, and which goes under
the name of *adenitis universalis.* You remember
when we were studying the initial lesion, I called your
attention to the induration of the chain of glands nearest to the lesion, and told you at the time how important it was. As the period arrives for the outbreak of the subsequent lesions, the glands all over
the body, the anterior and posterior cervical, the submaxillary and submental, the anterior and posterior
auricular, the occipital, the epitrochlear, and the inguinal glands are found enlarged and indurated. This
manifestation is coincident with the erythema cutis
et faucium, and with the alopecia which marks the
early stages of syphilis. Under treatment these indurated glands slowly subside, but their subsidence is
very gradual, and, if the result has been very good, no
trace is left behind ; but usually a slight hardness remains even after the patient has entirely recovered
from his illness, sufficient to show the practised finger
that trouble has existed.

This induration differs very widely from the brawniness and hardness which obtains with some chancroids. The condition of the glands in chancroid
you are already familiar with, but the adenitis in this
stage of syphilis you are not conversant with. In
the first place, the glands are painless; secondly, they
are unattended with acute inflammation; and thirdly,
they do not suppurate. They appear as round ker-
nels, from the size of a small buckshot to that of a
large pea, lying just beneath the skin, and upon
handling they roll about perfectly freely under the
tissues. This constitutes the form of infiltration of
glands which occurs during the early stages of syphilis; in the later stages of the disease another variety
occurs, which is entirely different in its course and
nature. This is called the gummous infiltration of
glands, and resembles in a slight degree a chancroidal bubo, just as the broken-down gummata of
the penis will simulate a chancroid. It begins as
an infiltration, not only of the gland itself, but of
the circumglandular tissue, which becomes tense
and brawny, and breaks down unless its course be
checked by proper treatment. There is one very
notable point in this breaking down: the skin
covering the swelling opens in several places, and
what comes from the enlargement is not pus, but
a thin, sticky, colorless fluid, not unlike thin gum.
This exudation is not abundant at any one given
time, but comes away continuously, and its dis-

charge does not materially diminish the size of the swelling.

This completes the circle of symptoms on the skin and mucous membranes likely to be met with in the average cases of syphilis which will fall to your lot, as practising physicians, to treat. But there are other lesions to which I wish to call your attention, fully as important as any you have heretofore studied, the consideration of which I shall reserve for a separate lecture.

LECTURE VI.

THE lesions we are to consider to-day are those which affect the special senses of sight, hearing, smell, and generation ; and as most of them occur in the late and more dangerous stages of syphilis, a correct knowledge of their natural history and course is important.

Commencing with the eyelids, we find that the skin and mucous membranes of these organs are sometimes the seat, during the early stage in syphilis, of the initial lesion and of mucous patches ; but as these symptoms do not differ in their general character from those found elsewhere upon the body, they need not detain us. During the later stages, the lids may also be attacked by pustules or gummata, which pursue the same course that similar lesions do elsewhere ; and the description which I have given in the two previous lectures will answer for these lesions of the lids.

When the initial lesion or mucous patches are seated upon the palpebral conjunctiva, some inflam-

4

mation of this tissue may ensue ; but it is usually very slight and limited in extent.

One of the most serious syphilitic lesions of the eye is what is known as iritis, or an inflammation of the iris ; and this is doubly dangerous because, from its close relation with the ocular vascular tunic—the choroid—the disease is liable to invade the deeper tissues and result in serious consequences to vision.

This symptom usually comes on about the sixth month of the duration of the syphilis—sometimes, however, as early as the third ; and is associated with a syphilide of the skin and mucous membranes. It commences with what the patient calls a "weakness of the eye," which, upon examination, is found to be very much congested, and this congestion is present, not only in the vessels of the conjunctiva, but of the sclerotic also. It is more marked close to the border of the iris, near the cornea, and is attended with lachrymation and photophobia. Upon close inspection, the iris of the affected eye is seen to be of a dull hazy color, to have lost its lustre, and it looks as though it were infiltrated with fluid. The pupil is small and contracted; and if a few drops of atropine be dropped into the eye, the opening will be found irregular in shape, and the pupillary margin of the iris bound down to the anterior capsule of the crystalline lens.

Note, then, these points in syphilitic iritis :

1st. *Congestion of the vessels of the conjunctiva and sclera.*

2*d. Lachrymation.*

3*d. Photophobia;* and

4*th. Adherence of the pupillary margin of the iris to the anterior capsule of the lens.*

In addition to these symptoms, the patient complains of a severe supraorbital pain, which, although present during the day, is more intense at night, depriving the patient of rest and sleep.

It seldom happens that both eyes are attacked simultaneously; the usual course is for one eye to be affected first; as the disease subsides in that, the second one succumbs, and upon its recovery the first one is a second time attacked—making what is known as a "see-saw iritis."

This is the variety which generally occurs in the early stage of syphilis; but later on, another kind appears, which is still more serious. The congestion, lachrymation, photophobia, and supra-orbital pain are again present in a more intensified form; the infiltration is more marked; and at the pupillary margin of the iris, apparently springing from the uvea iridis, an irregularly shaped nodule is seen which protrudes into the anterior chamber, sometimes completely blocking up the pupil. This nodule may break into the anterior chamber, and it then discharges a peculiar-looking flocculent fluid, which is not pus, but gummy matter. This form of gummous iritis is often conjoined with a pustular eruption upon the skin, or with gummata of some portion of the body.

Under proper care and treatment, the inflamma‧
tion and congestion subside, the iris assumes its normal color, and if the adhesions have not been very
firm, the pupil regains its normal contour ; but too
frequently the adhesions are permanent, and the pupil, particularly when dilated by atropine, shows an
irregular border   This may not, however, be a serious matter, nor does it necessarily affect the vision.

In the next stage, however, it is different.   As the
disease progresses, the deeper tissues are affected,
and the patient complains of dimness of vision,
and deep-seated pain in the eyes.   Examination
shows the normal range of vision diminished, and
the ophthalmoscope reveals infiltration of the choroid,
haziness of the choroidal and retinal vessels, with pig-
mentary deposits in the choroid, which are later on
succeeded by atrophy, leaving the sclerotic visible
beneath.

As might be expected, diminution of the range of
vision follows ; although in such cases you will be surprised to find how extensive are the ravages of the
disease, compared with the amount of loss of sight.

Besides these affections of the eyeball proper, the
carunculæ lachrymales and the lachrymal gland may
be the seat of gummata.   This is attended by swelling of the parts, which may, under treatment, disappear, or it may break down and leave an ulceration similar to other ulcerating gummata of the skin.

The *syphilitic affections of the ear* are not so well

understood as are those of other parts of the body. The auricle and external auditory canal may be the seat of mucous patches, and this variety of lesion belongs, of course, to an early stage. In addition, the middle ear may also be affected in the early as well as late stage, and this is due to a probable infiltration of the mucous membrane of the middle ear, as well as to an extension of the disease from the throat along the Eustachian tubes. The symptoms complained of are a feeling of tension in the ear; sometimes tinnitus aurium, although this is not constant ; and a diminished power of hearing. These are frequently associated with nocturnal hemicrania, and the early syphilides of the skin and mucous membranes. The speculum may show no trouble of the tympanum, or at the most a soggy condition of this membrane, with a slight sinking of the drum-head. This early lesion is not usually serious, as the symptoms pass off without affecting audition to any marked degree.

But, when syphilis invades the deep portions of the ear—the labyrinth and cochlea—then you may expect serious trouble, and the patient can consider himself lucky if he retains even a portion of his hearing. In such cases the symptoms are vague and ill-defined, being limited to pain in the head, which is not specially nocturnal in character, and occasionally tinnitus aurium. These continue for a longer or shorter time, when the patient suddenly wakes up some morning to find himself perfectly deaf. This peculiarity of suddenness in the attack is one worth your study, for you will find, when you come to examine other cases of nervous syphilis, that the same trait is present. The deafness is complete; the watch and tuning-fork, when pressed against the ear, convey no sound, and very often the same is true when these instruments are pressed against the bones of the skull or the teeth. I need hardly tell you in such cases the prognosis is not favorable. The cranial pain is frequently severe, and is not confined to any one portion of the head; sometimes being occipital, sometimes frontal, and at other times it is vertical or basilar. The tinnitus is the most distressing phenomenon in these cases, and is extremely rebellious to treatment, lasting even after a portion of the hearing power has been restored.

The *nose* and *air-passages*, in common with the rest of the body, are liable to invasion from this infernal disease, which spares no tissue of the human frame, but preys on all alike. In the early stage of syphilis the nasal mucous membrane becomes congested, and is the seat of mucous patches both in its anterior and posterior portions. These manifestations yield readily to treatment, and produce only slight discomfort; but as the disease progresses the parts are attacked by ulceration, with or without necrosis of the nasal and palatine bones, which give rise to a very fetid, abundant discharge. This is known as *ozæna syphilitica*—a form of ulceration so disgusing and offensive, as to render the subject of it a burden to himself and a curse to those who are brought in contact with him. If conjoined with necrosis of the nasal bones, the latter are stripped of their periosteum and crumble away, causing collapse of the bridge and sides of the nose, materially altering the appearance of the face. This stage of the disease is frequently associated with gummata elsewhere, either of the skin or mucous membranes.

The pharynx, the arches of the palate, the velum palati, and the mucous membrane of the hard palate, are, during the early period of syphilis, the seat of mucous patches, as well as of an erythema coincident with a similar affection of the skin. Besides these symptoms, later on in the disease, ulcerations, at first superficial, afterwards deep, occur, which are serious according to their extent and depth; but the most important lesion which can attack these regions is a gummous infiltration. This is grave in a twofold sense; first, from the impediment to respiration which the swelling gives rise to, and, secondly, from the after-effects which follow cicatrization of the ulcer. The first sign of this trouble is a feeling of fulness in the throat, with some embarrassment in breathing, due to the sometimes enormous swelling of the tissues of the part. This may be uni- or bilateral; when the latter, the impediment to respiration is very marked, and may necessitate a resort to tracheotomy, to relieve the urgent want of breath. This swelling goes on, unless checked by treatment, to ulceration; and the resulting sore is deep, with undermined edges, and a copious discharge of gummous matter and pus. If the velum palati be the seat of the lesion, perforation and absolute destruction of this septum may result, throwing the oral and posterior nasal cavities into one. When ulceration of the pharynx is present at the same time, the cicatrization which ensues produces a partial stenosis of the upper portion of the throat. One result sometimes occurs, of which I have shown you two examples, and it is this : when the soft palate is only partially destroyed, what remains becomes adherent to the posterior pharyngeal wall, producing occlusion of the entrance to the posterior nares, which would be complete but for cribriform openings in the artificial septum, through which nasal respiration is imperfectly carried on. If the perforation of the velum is limited in extent, under proper treatment the opening may contract to a size only sufficient to admit a very fine probe ; but my experience has taught me that *very rarely indeed* does the opening *entirely* close up. However, under favorable circumstances, the hole left behind is so small as to give rise to no trouble, nor to allow regurgitation of solids and liquids, such as obtains while the opening is large.

When the trachea and vocal cords are affected, the symptoms which follow are grave and alarming :

phonation above a hoarse whisper is prevented ; the tracheal rings are often necrosed and thrown off ; and death from suffocation may result from œdema and ulceration of the glottis.

The œsophagus is also invaded, usually in connection with syphilis of the larynx and trachea, either from an extension of the ulceration, or else from gummous infiltration of the tube itself. The stricture of the œsophagus, which results after cicatrization of the syphilitic ulceration, is a very grave complication, and frequently leads to a fatal termination from exhaustion, due to inanition, as solid food cannot be taken in sufficient quantity to support life.

The tongue, as we have already seen, is the seat of mucous patches during the early stage of syphilis, and these symptoms are often quite obstinate, recurring again and again when all other manifestations have apparently vanished. From being slight and superficial, the mucous patches may, during the progress of the disease, become painful and ulcerated, due in part to the·disease and in part to friction against the teeth. In the later stages of syphilis, the tongue may be attacked with a gummous infiltration, which may be diffuse or circumscribed. In the former variety, the entire substance of this organ becomes enormously swollen and thickened ; the surface is glazed, and presents deep and ulcerated fissures ; mastication is interfered with, and speech rendered indistinct. If the gummata are of the circumscribed

4*

form, one or more nodules, hard and cartilaginous to the touch, are felt deeply imbedded in the tissue of the organ. These nodules are painless, and do not occasion the patient as much inconvenience as when the lesion is diffuse. Both types may pursue two courses: resolution or ulceration. If the first, the thickening and ulceration gradually subside, the tongue regains its former pliancy, mastication and speech are recovered, and the organ shows no trace of its former trouble. When ulceration occurs, the discharge is apt to be abundant and ill-smelling ; the ulcer deep and excavated, surrounded with a thick margin of brawny infiltration ; mastication and speech imperfect, while the movements of the tongue are materially hindered. This ulceration sometimes lasts for months, causes great destruction of the organ, and when it finally heals up, leaves a puckered, depressed cicatrix, which may deprive the tongue of its accustomed mobility.

Passing to the *generative organs*, we find that the testicles are not infrequently attacked by syphilis in both the early and late stages. In the early period, the epididymis of one or both testes is hard, thickened, and distended to a sometimes enormous size. This enlargement is not painful, and only attracts attention from its weight and from the dragging sensation it produces upon the spermatic cord, causing a feeling of weakness in the back. This form of epididymitis almost always disappears under proper treatment, and does not interfere with the functions of the part.

This is not the case in the advanced stage ; here a true orchitis is found involving the entire organ. The first thing to attract the patient's attention is a sensation of weight in the part, accompanied by a dragging upon the spermatic cord, and a pain in the small of the back. Upon examination, the entire testis is found very much enlarged, hard as a stone, and presenting upon its surface raised projections or knobs. There is no redness, and, curious to tell, *no pain;* the organ can be very freely handled without exciting any uneasiness. This peculiarity is also present in syphilitic epididymitis, and in this respect it differs very much from the gonorrhœal form of this disease, as I shall show you by and by. The shape is piriform, with the small end pointing towards the abdominal ring. If the disease pursues a favorable course, the hardness and infiltration subside, and the organ may return to its former size and usefulness ; but too often atrophy results, and the testis all but disappears, sometimes being no larger than a good-sized horse-bean. Of course, when this happens, it is neither ornamental nor useful.

The other course which the disease may pursue is ulceration. One or more of the projections soften, break down, and discharge a mixture of pus and the gummous material with which you are already familiar. The ulcer differs in no respect from broken-

down gummata elsewhere ; is chronic, sometimes lasting for months before it finally heals up ; and when it
does, leaves behind it a deep, depressed scar, surrounded by atrophied tissue, which is not so extensive as where the infiltration has been more general.

Gummous infiltration may also occur in the *ovaries;*
and the only symptom present is swelling, usually
painless, in the ovarian region, conjoined perhaps
with some symptom of syphilis elsewhere.

The *cervix uteri* is not infrequently the seat of the
initial lesion and of mucous patches, and these symptoms we have already studied in a previous lecture.
But there is one point in this connection to which I
invite attention.   Both these lesions may be seated
within the cervix, between the os internum and exter-
num, showing nothing externally ; a slight discharge
is present, but no more than is common to nine
women in ten.   Connection with women thus affected gives rise, in the male, to an attack of syphilis,
and, unless care is taken in forming the diagnosis,
may occasion the error of regarding the syphilis of the
man as arising from a gonorrhœa or leucorrhœa in
the female.

This would be an error ; the disease is contracted
from the secretion of an initial lesion or of a mucous
patch ; indeed, accept this axiom : *syphilis comes only
from syphilis, and not from clap or a chancroid.*   In
the advanced stage of syphilis the neck and the body
of the uterus are attacked by gummata, which pre-

sent themselves in the shape of diffused or circumscribed thickening of the organ, which may follow
the usual course of these lesions ; viz., absorption, or,
as sometimes happens in the cervix, deep and obstinate ulceration, resembling, in many respects, an extensive chancroid of the part. We also find the same
in the male, seated upon the penis, to which I have
already alluded in a previous lecture. This gumma,
besides occurring upon the mucous membrane of the
male genital organ, is sometimes found at the junction
of the penis with the scrotum—the peno-scrotal angle
—as a hard, diffused, brawny swelling, unattended by
pain or redness ; this opens externally or internally,
and according as it does one or the other, gives rise
to certain symptoms. If the opening is external, the
resulting ulcer is similar to the ulcerating gummata
of other parts ; and heals up, after a longer or shorter
time, under appropriate treatment. If, on the other
hand, the urethra is perforated, the gumma discharges itself through this canal, and gives rise to the
question of gonorrhœa with a peri-urethral abscess.
In the majority of cases you will save yourselves from
falling into the error of regarding this lesion as a clap,
by an observance of the following facts : in gonorrhœa the discharge precedes the swelling, which is
red and *painful ;* in syphilis, on the other hand, the
discharge *follows* the appearance of the enlargement,
never precedes it, and the swelling is *neither red nor
painful.*

Another form of gumma of the male genital occurs, to wit : an infiltration into the corpus spongiosum or into the corpora cavernosa. This usually
comes on in the circumscribed form, and is apparent
as a hard nodule without redness or pain, deeply
imbedded in the tissue of the part, or it may occasionally present an annular form round the entire
organ. It generally passes off under treatment, but
during its continuance it gives rise to much inconvenience and to most curious distortion of the part.
If only partial, it curves the penis during erection to
one side or the other, according to the location of the
gumma, resembling the symptom known in gonorrhœa as chordee, and it interferes with sexual intercourse ; but if it assume the annular form, a most re-
markable condition of affairs arises. During erection,
the penis, from the crura to the seat of the lesion,
is turgid, and assumes its usual appearance ; beyond
that, it is flaccid and hangs at right angles to the rest
of the organ, looking like a flail. Of course, for
sexual purposes it is entirely useless, and but for the
glory of the thing, the poor patient might as well
have no penis at all.

The *alimentary canal and the viscera* do not escape
any more than do other portions of the body. During the existence of the erythema, a form of *icterus*
has been described as due to syphilis, which yields to
mercury, but it is not until the later phases of the
disease that these organs are attacked by gummata,

usually of the circumscribed variety. These lesions have been found in the liver, lungs, heart, kidneys, and intestinal tract; but the most interesting of all the syphilitic manifestations of these parts is the gumma of the rectum. This begins in the muscular and mucous coats of the rectum, and by its size may decidedly diminish the calibre of the tube. The neoplasm ulcerates, producing great pain, attended with a discharge of purulent and gummous material— tenesmus, diarrhœa, and bloody stools. Upon healing, it leaves behind it a stricture of the rectum, which is more or less tight, according to the depth and extent of the ulceration, and is usually attended with obstinate constipation from the mechanical obstruction to defecation. This stricture is extremely obstinate and rebellious to treatment, owing to its continual irritation by fæcal matter, and necessitates a resort to the use of rectal bougies to keep the passage dilated, and even surgical interference, such as a division of stricture, or even in extreme cases to colotomy, to prevent the rectum from being occluded and the patient's life jeopardized.

This lesion of syphilis also is of interest in its bearing upon the chancroid. You remember, perhaps, that in the lecture upon chancroid, I spoke of the stricture of the rectum resulting from anal chancroids in the female. The after-effects differ in no whit from the same disease due to syphilis, and may also require subsequent surgical treatment for its relief.

LECTURE VII.

SYPHILIS OF THE NERVOUS SYSTEM AND OF BONE.

THUS far we have studied the syphilitic lesions which occur in and upon the body, with the exception of those which affect the nervous and osseous systems, and these we shall study in to-day's lecture.

It is generally believed that nervous symptoms belong exclusively to the late or so-called tertiary form of syphilis, but this is a mistake ; lesions of the nervous system are found during the early period, being sometimes coincident with as early a manifestation as erythema, but they differ from late nerve-syphilis in being evanescent, more amenable to treatment, and in not leaving any permanent impairment of the health behind.

One of the most common symptoms of the early stage is the hemicrania or headache confined to one lateral half of the head, and to which, when speaking of the syphilides of the skin, I called your attention. This headache has one peculiarity, especially well marked in the early period of syphilis; it only appears at night; during the daytime the patient is free from it, but on the approach of night it com·

mences gradually at first, increasing in intensity, when the patient goes to bed, and remaining until morning, when it disappears. It usually affects one lateral half of the head, although it may shift its position to the frontal and occipital portions; but this is not common. As the syphilis advances, this nocturnal character changes; it no longer disappears entirely throughout the day, although it is less severe in the forenoon. Some time in the afternoon it begins to increase, and at night becomes so intense as to deprive the patient of rest and sleep. The more severe and advanced the type of the syphilis is, the earlier in the afternoon does this pain commence.

Associated with this hemicrania are epileptiform seizures of a light and transient character, which, so far as the patients are concerned, pass unnoticed, for the simple reason that they know nothing about them. Sometimes an attack occurs in public, when of course it becomes known; but at other times the only things to excite suspicion toward such a manifestation are a bitten tongue or lips, or else a bruised forehead. I recall one case of a young interne of this hospital (Charity, B. I.), who contracted syphilis from a wound on the finger becoming inoculated, and who had these epileptiform convulsions during the existence of his erythema. He had repeated attacks, and in many of them was seen by his room-mate and by other internes of the hospital. My then partner, the late Dr. Bumstead, and myself were visiting in

the wards, and the case came under our care. The
attacks were of short duration, but well marked, the
jaws were firmly set, there was no biting of the
tongue, and but slight foaming at the mouth ; the
body was first rigid, then violently convulsed for a
minute or more, and all was over. The patient re-
mained dazed for a few minutes after, would then
pick himself up from where he happened to be lying,
and finish what he was doing when the attack came
on. The most curious part of the attack was, that
the being on the floor or bed never struck him as
strange, and he seemed to be absolutely ignorant of
his attack. He entirely recovered under proper
treatment.

As the syphilis advances, these attacks become
more frequent and severe, and, unless checked by
treatment, affect the patient's mind, leading to an attack of downright mania, or, what is more commonly
the case, to melancholia and idiocy. But one point
is deserving of notice — the rapid and beneficial
effect which accrues in cases which at first look
almost hopeless, under a proper and thorough treatment.

Associated with these cases of syphilitic epilepsy,
although not necessarily so, are paraplegia and hemi-
plegia, attended with certain symptoms which serve
to distinguish them from similar affections not due to
syphilis. First and foremost of these, stands the suddenness with which the attack comes on ; the patient,

to use a slang phrase, is "bowled over" without pre-monition. Occasionally the patient will confess to having suffered for a short time before the attack with severe cranial pain, but just as often as not there are no antecedent symptoms ; the patient becomes sud-denly paralyzed. The second noteworthy point is, that *very rarely* indeed is there any loss of consciousness ; the patient retains his senses perfectly, has neither stertor nor coma ; he simply finds he cannot move certain portions of his body. If he is attacked with hemiplegia, one lateral half of his body is useless—if paraplegia, the lower half ; and this latter form is con-nected with obstinate constipation and with retention of urine, from the inability of the rectum and bladder to empty themselves of their contents. Paraplegia denotes some affection of the spinal cord, low down, as a rule, and due to compression either from the pressure of a gumma in the periosteum of the verte-bræ, or in the sheath of the cord itself ; while hemi-plegia is caused by some brain lesion.

Age plays a part also in making up your diagno-sis ; and you will remember that such lesions as we are now considering, occurring in an adult say be-tween the ages of twenty and forty-five, of course excluding accidents, should always excite a suspi-cion of syphilis ; for, apart from injuries and the pox, these diseases are rare between the ages I have given you.

Let me supply you with a short table of the differential signs between syphilitic and non-syphilitic paralysis :

SYPHILITIC PARALYSIS.	NON-SYPHILITIC PARALYSIS.
Sudden, unattended by premonitory symptoms.	Gradual, and attended by prodromata, except in apoplexy, when the
Consciousness not lost.	Patient becomes unconscious.
Breathing calm, no stertor.	Breathing stertorous.
Pulse regular and natural.	Pulse full, bounding, and irregular.
Most common between the ages of twenty to forty-five.	Usual in advanced age.

This tabular form will, I hope, serve to fix these points in your mind.

Syphilis of all diseases seems fond of playing curious pranks, and the nervous system affords it ample opportunities. Besides the varieties of paralysis which we have just gone over, there are localized forms that attack certain muscles or sets of muscles. The most common of these is paralysis of the muscles supplied by the third pair of nerves, the motores oculorum communes. In this affection the eye-ball is partially or completely covered by the lid, which cannot be raised, and the eyeball is incapable of any movements except those afforded by the external rectus and the superior oblique muscles, which you know are supplied by the fourth and sixth pair of nerves. This produces disturbance of sight, with diplopia or double vision, from inability to focus the two eyes simultaneously upon the same object. It also affects the iris, producing mydriasis or dilatation of the pupil, which is sometimes extreme.

Next in frequency come the affections of the fifth and seventh pairs, and here we find a complete distortion of the muscles of the face supplied by this nerve ; the face is pulled over to the *non-paralyzed* side, because there are no antagonistic muscles in action to keep the features straight. The tongue, when protruded, is dragged over in the same manner. The patient cannot inflate his cheeks, nor can he masticate his food, as the buccinator and masseter. muscles are both incapacitated ; his food collects during eating between his cheeks and jaws, and cannot be dislodged save with his fingers, and the saliva dribbles out of the corner of his mouth. He presents, in short, a ridiculous and at the same time a pitiable appearance. Besides this he cannot close the eyelid of the affected side, and as for winking with it, that is out of the question ; the ala nasi of that side does not expand in respiration ; he cannot wrinkle the skin of his forehead, nor can he frown but with one-half of his face, and he may also be made deaf on the diseased side. Yet with all this trouble, if the fifth is not attacked, there is no loss of sensation, for the seventh, as you know, is the motor, while the fifth is the sensory nerve of the face. Whether all these symptoms or only a portion of them occur, depends upon the site of the lesion ; if it is anterior to the emergence of the nerve through the stylo-mastoid foramen, all are present ; if posterior, then only those muscles supplied by the diseased portions of the nerve are affected.

If the fourth pair is attacked, then the obliquus superior is the only muscle at fault, and the patient cannot turn the eyeball upwards and outwards, and if the sixth pair is injured the eye cannot be everted.

It is so rare to find these forms of localized paralysis apart from syphilis, that I do not believe you will ever be far wrong in ascribing such lesions to this disease ; and in cases where no history can be obtained, the importance of a knowledge of this fact will be at once apparent to you.

Let me, then, formulate this into an axiom for you :

Paralysis of single muscles, or sets of muscles, are nine times in ten syphilitic.

These affections of the nerves are nearly always unilateral, and I do not know that they occur more frequently upon one side than the other.

Among the spinal nerves, the one most commonly attacked is the great-sciatic, which springs from the sacral plexus. The principal symptom present is pain along the course of the nerve, and this pain is not acute, but dull and persistent, and is liable to exacerbations at night. None of the ordinary remedies used for sciatica do more than mitigate the severity of the pain ; but if the surgeon gets upon the right track and prescribes the iodide of potassium either alone, or, better still, combined with mercury, the result is oftentimes as rapid as it is gratifying; the pain vanishes like magic.

The lesion which occurs in these nervous syphilides is twofold; either a deposit of gummous material within the nerve-sheath itself, or else pressure upon the nerve during its passage through some bony canal or foramen by gummata of the bone. The prognosis depends much upon the duration of the disease; if the syphilis be young—*i. e.*, in its early stage—it is favorable; if the contrary, the prognosis is doubtful, although even here hope should not be abandoned; but if atrophy of the nerve has resulted from the pressure of the gumma, then good-by to all chance of recovery.

As regards the *bones*, the lesions here are divisible into those which occur during the early and those which occur during the late stages. To the former belong the osteocopic pains, which produce no organic changes in the bones themselves nor in their investing sheath the periosteum—which are nocturnal in their character, and are at the worst merely annoying. As the syphilis progresses, these pains lose a great deal of their nocturnal character; they are more persistent, but still, with all this, they are not dangerous. These pains are usually confined to the shafts of the long bones, particularly those which are just beneath the skin, such as the tibia and the ulna; although they sometimes affect the flat bones—as, for example, the cranial.

It is when the gummous stage arrives, that trouble of a serious nature arises. The first stage is where intense localized pain occurs in some bone either flat or long—it makes no difference—which is speedily followed by a swelling at this spot, oftentimes exquisitely tender, but usually without any redness of the part. This swelling, if checked at the outset, disappears slowly, nearly always leaving some elevation and thickening of the periosteum behind it. If left to itself, or uncontrolled by the treatment, the swelling increases in size and extent, gradually softens and opens in one or more places to give exit to pus and the gummous material which is common to all the lesions of the late stage. If this opening be probed, dead bone is almost always found at the bottom, and this bone conveys to the touch an irregularity on the surface as though it were worm-eaten. And here let me impress upon your minds one very important maxim : NEVER, *never under any circumstances, open a gummous enlargement of bone or gland, no matter how soft it gets.* I have seen gummous infiltrations of this kind become absorbed even when the skin covering them was as thin as fine tissue-paper, and they looked as though they must open. I say to you again, never open a gumma, for by so doing you deprive yourself of the only chance of preventing necrosis of the bone ; and if this must supervene, do not give it a helping hand by stupid interference on your part.

But we will suppose necrosis already present ; what happens then ? The tumor keeps on discharging, and in the discharge fragments of crumbling bone are found. Let me say that the extent of the necrosis is usually confined to the size of the periosteal swelling, so that when death of the bone has once set in you can have some idea of its limit. The bone crumbles away little by little, presenting nothing in the shape of a firm sequestrum for you to extract ; indeed, it seldom has the line of separation from sound bone which dead bone of non-syphilitic origin shows, but it simply chips off in small flakes and pieces until it has reached the limits of the diseased portion, when, if treatment has been properly pursued, it stops, granulations spring up from the bottom and sides of the cavity, cicatrization takes place, and a more or less depressed cicatrix is left behind to mark the loss of bone.

When this necrosis occurs in the external osseous framework, the results, although bad, are seldom serious ; but when it occurs in the internal bones, such as the palatine, nasal and hyoid, or in the rings of the trachea—for cartilage disappears as well as bone —then serious mischief follows, not confined alone to the shocking disfigurement which occurs, but it may even endanger the patient's life. The same process is repeated here as in the long bones ; the gummous deposit takes place into and beneath the periosteum, stripping the latter from the bone ; necrosis and exfoliation of bone follows, and when this happens in the palatine and nasal bones the oral and nasal cavities are thrown into one, and the disease may go so far as to attack the base of the skull, causing coma, low delirium, and death. These are the cases so frequently associated with syphilitic cachexia ; and when that stage is reached, hope is about at an end. You may perhaps recall such a case which I showed you from ward 13, where the hard and soft palate had both disappeared, the nasal bones had gone, causing the nose to flatten out upon the face ; where necrosis of the vertebræ at the posterior pharyngeal wall was present, and a sinus led from the inferior orbital angle to a mass of dead bone in the lower plate of the orbit. I called your attention to the condition of the man, and to his worn-out, more dead-than-alive look, and told you then his race was nearly run. He died a week after, in spite of treatment, gradually sinking into a low form of delirium until death released him from his sufferings.

These are the cases, happily rare, which once in a while present themselves as if to show what syphilis is capable of doing, and there is one more form about which I wish to speak to you before closing this lecture. This is where syphilis attacks the rings of the trachea, and where, from pressure of the gumma upon the glottis and trachea, death by suffocation threatens to supervene, rendering tracheotomy necessary to save life. Under active and persistent treatment the neoplasm may disappear, but too often the cartilage exfoliates ; the rings disappear, and upon cicatrization a partial stenosis of the trachea occurs ; and this impediment to respiration, combined with the exhaustion so often found in these cases, rarely fails sooner or later to end the patient's life.

The *tendons* also participate in this disease, and are usually attacked in the late stages by a gummous deposit in. their sheaths. While this lasts it may produce curious deformities ; as, for example, when it occurs in the tendo-Achillis, it produces a talipes equinus, and if in the tendons of the flexor communis digitorum it imparts to the hand a peculiar claw-like look. Of course such a hand is useless.

The symptoms are those of gummata elsewhere, swelling and thickening of the parts, unattended by much pain. They usually yield to treatment, but sometimes permanent contraction ensues, rendering tenotomy necessary in order to restore the parts to some degree of usefulness.

We have now run over the principal points in the history and course of syphilis, and I trust that the pictures I have sketched for you in these lectures will enable you to recognize all the cases which you will be likely to see in every-day practice. The next lecture will be devoted to the treatment of syphilis— a very interesting subject, and to the importance of which I think you are keenly alive.

LECTURE VIII.

As regards the treatment of syphilis, allow me to say at the outset that it would not come within the scope of these lectures to discuss *pro* or *con* the various methods which have been in vogue since syphilis has been recognized as a separate disease; and what I therefore propose to do is, to give you the kind of treatment which has best stood the test of time, and which at the present day is the most approved. With this object in view I shall divide the subject into the two principal groups of internal and external treatment, and give you as I go along the appropriate prescriptions for each.

In the first place, as regards the treatment of the *initial lesion.* I have already, when speaking of this form of syphilis in a previous lecture, given you the plan most deserving of adoption, and will therefore do no more than refresh your memory upon some of the principal points to which I then called your attention.

In the first place, *do not cauterize* the initial lesion unless it be attacked by phagedena, when it may be admissible ; but when it is uncomplicated, cauterizing it *does no good;* on the contrary, it *does harm.* In the second place, do not treat it by the internal use of mercury, for the reason that this metal retards the appearance of the early syphilides, and leaves the surgeon at sea as to when to expect subsequent lesions, and what to look for, and also because its use sometimes prevents the surgeon from deciding with certainty upon the nature of doubtful ulcers ; and when, the period of probation passing by, and no symptoms appearing, he assures his patient that nothing further is to be expected, his promises of future indemnity are apt to be rudely dispelled by the appearance of the long-delayed syphilides some months later. In addition to this, waiting until the syphilides appear does not injure the patient's chances of ultimate recovery. Treat the initial lesion, then, by the rules laid down in Lecture III.

When the syphilides appear, however, and the time for internal medication arrives, what shall we do ? In the early stages of syphilis, you remember, the symptoms are multiple and polymorphous; and when the six weeks of incubation have elapsed, your patient blazes out with an erythema of skin and mucous membranes, papules in the scalp, mucous patches of the tongue and throat, alopecia, hemicrania, and universal induration of the glands of the body. Preceding these symptoms there probably has been some febrile excitement, which disappears as the eruption shows itself. Now is the time for the use of mercury; and let me tell you that, of all the drugs at your command for the treatment of syphilis, there is not one that will take its place. Dismiss from your minds the senseless abuse of mercury which some writers indulge in, and remember that the surgeon who neglects to use this mineral in treating syphilis does injustice both to his patient and to himself; for although some mild cases of syphilis may and do recover without its use, the risk run is greater than any prudent surgeon should incur. Know what to expect from your drug, use it properly, and depend upon it that those two points well carried out, the mercury will do no harm either in the present or the future; on the contrary, it will do good.

In the early stages of syphilis, *i. e.*, through the period of erythemata and papulæ, the preparation that I use, I was going to say to the exclusion of almost everything else, is the following:

Ŗ.   Mass. hydrargyri .... gr. ij.
      Ferri sulphatis exsiccat............ gr. i.
      Fiat pil. no. i.
M.
Sig.—Three to six daily.

I usually begin with one three times daily, after meals, gradually increasing the number to two, three times daily, as occasion requires.

The bichloride of mercury is the old and time-hon‧ored preparation which has been usually given. I very seldom use it, because in my hands it has been apt to produce its toxical qualities, griping of the bowels, diarrhœa, and sponginess of the gums, just when it is most needed. Still, in some cases it answers well enough, and when used it had better be given in pill form, thus :

   ℞.   Hydrargyri bichloridi .... gr. $\frac{1}{16}$–$\frac{1}{8}$
        Saponis...................... q. s.
        Ut fiat pil. una.
  Sig.—One, thrice daily after meals.

In order to check its action upon the bowels, from $\frac{1}{4}$–$\frac{1}{2}$ grain of opium may be added to each pill.

Another form, one much esteemed by Ricord, is the

PROTIODIDE OF MERCURY PILL.

   ℞.   Hydrargyri protiodidi .... gr. $\frac{1}{3}$–$\frac{1}{2}$
        Extracti gentianæ.............. q. s.
        Ut fiat pil. no. i.
  Sig.—One, thrice daily after meals.

But of all these preparations of mercury, as already stated, I much prefer the one first given, the blue mass and iron pill, for its efficacy and for the tolerance which the system shows to it. The addition of the iron is of value not only in increasing the action of the mercury, but for its own effect as a tonic.

Now comes the question : how long shall the mercury be continued ; how much shall be given ; and under what circumstances shall it be increased, di. minished, or stopped altogether ? To the first point I reply, until the symptoms disappear or the drug produces toxical symptoms ; by that I mean disturbance of the digestion, diarrhœa, sponginess of the gums, and salivation. With regard to this last, I wish to impress upon your minds the fact that its occurrence is a hindrance, not a benefit to treatment, inasmuch as, when present, the mercurial has to be stopped and so much time wasted. *Avoid then, carefully, any approach to salivation ;* but should such an accident occur, suspend all antisyphilitic treatment and place your patient upon the following prescription :

R . 　Potassæ chlorat...................... ℥ i.
　　　Aquæ............................ ℥ vi.
M.
Sig.—Locally as a mouth-wash, and internally in teaspoonful doses, four or five times daily.

This checks the sponginess of the gums, the fetor of the breath, and the flow of the saliva, which are the three symptoms attending this form of mercurial intoxication.

Two other remedies have been used, both of which may be of service. They are belladonna, or its alkaloid atropine, and dilute nitric acid. They are usually given as follows:

    ℞.   Tinct. belladonnæ................. ℨ iv.

        Aquæ .... ℥ ij.

   M.

   Sig.—Teaspoonful four times daily, in water.

If you use atropine instead of belladonna, give the following:

    ℞.   Atropiæ.................... gr. $\frac{1}{10}$

        Alcoholis.................. ℥ ss.

        Aquæ..................... q. s. ad ℥ ij.

   M.

   Sig.—Teaspoonful three or four times daily.

With the preparations of belladonna use the solution of the chlorate of potash given above as a wash. The dilute nitric acid you will oftentimes find of benefit in those cases where the sponginess of the gums is so excessive as to threaten the dropping out of the teeth, and should be given both internally and locally.

    ℞.   Ac. nit. dil....................... ℨ iv.

        Aquæ........................... ℥ ij.

   Sig.—Teaspoonful four times daily, in water; also use locally.

If, however, you give mercury prudently and properly, carefully watching your patient, no such

  5*

accident as I have just detailed need occur; and, indeed, you will oftentimes be surprised to see how
tolerant the system is in syphilis of even large doses
of this mineral. I have often given in these early
stages of the disease 10 to 12, and even 14 grains a
day for several weeks at a time, without producing
any systemic disturbance whatever; but it was in
those cases where the attack was severe, and I was
careful to keep the patient under rigid observation.
In average cases, 6 to 8 grains daily will be sufficient
to dispel the symptoms.

As to the circumstances which shall impel us to
increase, diminish, or altogether stop the mercurial,
they may be disposed of in a few words. If the
symptoms be obstinate and slow to disappear, and if,
at the same time, the patient stands his treatment
well, the drug may be gradually increased until the
symptoms give way or the patient begins to show a
slight red line at the edges of the gums. Should
this latter occur before the disappearance of the
syphilitic lesions, the mercury must be suspended for
a few days, and when it is recommenced, a different
preparation given from the one formerly used. It is
seldom, however, that the earlier manifestations re-
sist a determined attack with this mineral.

As soon as the symptoms have disappeared so as
to leave no staining of the skin or other trace of their
presence behind them, it is well to discontinue the
use of the mercurial for the following reasons: first,

to avoid too great a tolerance of the system to the drug; and secondly, to enable us to determine whether other lesions are about to follow or not. Upon this last point let me dilate a little, even at the risk of seeming tedious, in order to avoid misunderstanding upon your part. We will take, for example, one of the many cases which I have already shown you from the wards—say this one, of a papular syphilide. As soon as the manifestations have disappeared from the skin, leaving no trace behind them, the mercurial treatment will be discontinued and the man will be placed upon tonics. Now, if you will remember what I have told you when we were speaking of the syphilides of the skin, you will recollect that there is a period of incubation, shorter or longer as the case may be, between the appearance of the various manifestations, and if you continue your treatment after the first train of symptoms have disappeared; you delay the occurrence of the subsequent ones. But suppose you intermit your treatment instead of continuing it, and the period of probation passes without the expected symptoms appearing—it shows you that the disease is losing its strength (for the amount of mercury you have already given for previous symptoms would not prevent the subsequent manifestations if the syphilis were still very active), and you would be justified in supposing that the disease was on the wane, and the longer the time which elapses between the various stages, the more hopeful the prognosis. *But bear in mind that as long as any symptoms last, no matter how slight, so long must the treatment be continued; and also that it must be renewed, if previously discontinued, should fresh manifestations recur.*

This touches upon internal treatment only; but occasionally some lesions require a topical as well as a constitutional medication. Of these, mucous patches head the list. The early lesions of the skin, of course, require no local applications; it is only where the erythematous blotches and papules invade skin and mucous membrane together, as, for example, at the angle of the mouth and eyelids, or in other portions of the body which combine heat and moisture, such as the pourtour of the anus, the labiæ vulvæ, the scrotum and penis, the toes, the buttocks and arm-pits, that topical treatment becomes requisite. The two best remedies for these lesions are powdered calomel and the application of the nitrate of silver, either in the solid stick or as a saturated solution. But do not forget the most important point of all : *keep the parts dry and clean*, else your treatment will be of little avail.

When the mucous patches are seated in the throat, or on the lips, tongue, and lining membrane of the cheeks, the application of the nitrate of silver is generally the most efficacious ; and when the lesions are seated low down in the pharynx a spray of a weak solution of nitrate of silver (gr. v. to aq. ℥ i.) will be of advantage.

As the later stages of the disease are reached, the treatment undergoes certain modifications; the one best calculated to promote a cure is that known as the *mixed treatment*. This consists of mercury and the iodide of potassium, used either separately or in combination, and is given in those stages of the disease which are ulcerative in their character. I much prefer giving the two separately, for facility of exhibition, and because either one can be increased without increasing the other. The two preparations of mercury most in use are the protiodide (internally), and the ordinary mercurial ointment, or oleate of mercury, as an inunction to the skin. If the internal use of the drug be decided upon, the protiodide should be given once daily, in from a half to one grain, and the iodide of potassium in two daily doses, thus :

R. Hydrarg. protiod.............. gr. ½–gr. i.
Ext. gentianæ................ q. s.
Ut ft. pil. una.
Sig.—Once daily, after mid-day meal.
And—

R. Kali iodidi .... ʒ ij.
Tinct. cinchonæ comp.,
Tinct. gentianæ.................āā ℥ ss.
Aquæ.....................q. s. ad ℥ ij.
M.
Sig.—Teaspoonful well diluted with water twice daily, morning and evening, after meals.

Should you elect to combine the mercurial and the iodide of potassium in one dose, you will find the following prescription a good one :

    ℞.   Hydrarg. bichlor................ gr. ¾–i.

Or—

        Hydrarg. biniodidi.............. gr. ¾–i.
        Kali iodidi.................... ℥ ij.
        Tinct. gentianæ,
        Aquæ......................āā ℥ i.

    M.

Sig.—Teaspoonful well diluted with water twice daily, morning and evening, after meals.

All of these remedies should be given after eating in preference to before, because the iodide of potassium sometimes produces intestinal disturbance if given upon an empty stomach.

But, we will suppose you do not wish to give mercury internally by the mouth, but prefer some other mode of administration. What ways are open to you ? There are three: first, by inunction, *i.e.*, friction on the skin, of some oleaginous or fatty preparation containing mercury; second, by mercurial vapor-baths; and third, by subcutaneous injections.

The first of these methods, by inunction, although a most excellent way of getting a rapid and at the same time thorough effect of mercury, is open to the serious objection of uncleanliness, and with justice, as the old-fashioned way of smearing the ointment over the entire body in divided doses kept the body and linen in a constant state of greasiness and dirt. This, in recent times, has been much improved upon by the use of the oleate of mercury; but this, though better than the unguent. hydrarg. of the pharmacopœia, is repugnant to many persons who are careful about the cleanliness of their persons. To obviate this, and to reduce the dirty feeling which any greasy substances impart to the skin, I have for some time past used the oleate of mercury, 20% strength, on the soles of the feet, to the exclusion of the ordinary mercurial ointment, in the following manner:

The patient is directed to bathe the feet thoroughly in hot water the night on which the first inunction is made, when half a drachm of the 20% oleate of mercury is rubbed briskly into the sole of the right foot; this is repeated the next night on the left foot, and so on alternate nights the right and left foot is anointed with half a drachm of the preparation. This may be increased to a drachm, or more, if the patient stands the mercurial well. The same stockings, which should be of wool or some tolerably thick material, are worn continuously, night and day, for one week, at the expiration of which time the feet may be thoroughly cleansed with hot water and soap, and an intermission of three or four days elapse before renewing this same process for a similar length of time. The iodide of potassium should be kept up during the period of inunction as well as during the intermission, in three daily doses.

The advantage of this method is twofold: first, as regards cleanliness; second, as to efficacy. Instead of smearing the body all over and keeping it in a continually dirty state, the whole of this disagreeable feature of the treatment is confined to the feet, and the repeated dose is in a process of continual absorption, inasmuch as every movement that the patient makes in walking serves to rub the ointment into the skin of the feet, and absorption does take place notwithstanding the thickness of the cuticle in this part.

The second method, by the vapor bath, is equally efficacious, and not open to the same objections that the inunction process is. The patient may be sent to one of the regular establishments where these baths are given, or, if preferred, it may be given in the patient's own house. The apparatus requisite for the purpose is the portable vapor-bath, sold in the shops under the name of " Lee's Mercurial Vapor Bath," or the American modification of the same—a long, sleeveless flannel night-shirt, made to reach to the patient's feet, and an India-rubber mackintosh of the same pattern as the flannel shirt, both of which should close tightly round the neck, leaving the head exposed, and a round stool for the patient to sit upon. The flannel shirt and the mackintosh should be made large enough to allow the patient to sit upon the stool inside of both. The vapor-bath is a cylinder of tin or of wire gauze, enclosing within it an alcohol lamp. The upper portion of the cylinder holds a plate, which is hollowed out in the shape of a gutter at its outer circumference ; the middle portion is elevated above this gutter, and contains a shallow depression or cup. The patient, being stripped and dressed in his shirt and mackintosh, is seated upon the stool which is included within his bath-clothing, and the whole is carefully tucked in at the bottom, to prevent the escape of any vapor. The bath is prepared in the following manner: water is poured into the gutter of the plate at the upper portion of the cylinder, and the mercurial is placed on the shallow cup at the apex, in the middle ; the lamp is then lighted, and the whole apparatus placed under the stool upon which the patient is sitting. The lamp is so arranged that the flame striking against the plate at the top causes evaporation of the water, and the heat throws the patient into a profuse perspiration, producing a steam vapor-bath. As soon as the water has evaporated, the mercury, in its turn, is volatilized and readily absorbed by the skin. As soon as all the mercury has disappeared, the light is put out, and the patient is left inside his waterproof clothing until the body begins to cool slightly; he should then be taken from his stool, the waterproof cloak removed, while the flannel shirt is retained, and he should be covered up with blankets until all perspiration has ceased and the body has become cool and tolerably dry, when he may put his clothes on again. This is supposing the bath to be given in the day, but bedtime is the best period of administration, when the patient may go to bed at once and remain there.

A good substitute for the lamp is an ordinary chafing-dish, the tin or zinc plate of which may be replaced by an iron saucer to contain the water, and this, upon evaporation of the water, becomes thoroughly heated. When this is accomplished, the mercury may be placed upon the still hot plate, producing the same result which is attained by the regular apparatus.

The preparation of mercury used is either calomel or the black oxide, the former being given in 20 to 40 grains to each bath, and the latter from 30 to 60.

The time required for the bath varies from thirty to forty minutes, and, barring the length of time it takes, is one of the nicest and cleanest ways of introducing mercury into the system, besides being of easy application.

The method by subcutaneous injection is very little used in private practice, owing to the trouble of administration and the pain attendant upon it. It is done by injecting the solution containing mercury beneath the skin, which, besides being painful, is frequently followed by abscesses at the point of injection. Calomel is the agent usually selected, and is given in doses varying from $\frac{1}{10}$ to $\frac{1}{2}$ a grain at each injection.

The local treatment of the ulcerative syphilides, although not so important as the constitutional treatment, is decidedly necessary and useful. Those of the skin, if the crust has been removed, should be dressed with mercurial ointment spread upon a cloth. It is better, however, to leave the crust on, if it be firmly adherent, as it makes the best protection for the part, and the underlying ulcer heals up under the administration of the mercury and iodide of potassium. The ulcerations occurring in the throat and mouth should be treated with nitrate of silver (40 grains to one fluid ounce of water), carbolic acid (crystals gr. xxx. to water ʒ i.) or nitric acid (ac. nit. c. p. ℥ xxx. to water ʒ i.). If the lesions are deeply seated in the throat, or in the posterior nasal cavity, they may be reached by a spray of the above solutions.

In necrosis of the nasal and palatine bones the parts should be thoroughly washed out with warm water, injected through a posterior nasal syringe, and afterward sprayed with the solutions given above, and patience exercised until the dead bone comes away under internal treatment.

I now wish to say a few words to you with regard to the administration of your remedies, because upon the thoroughness with which you use them will the advantage of your treatment largely depend. Without at all advising you to be rash, I wish you to be bold, and to remember that in face of such a disease as syphilis you cannot afford to trifle. When using mercury, watch your patient carefully, be on the lookout for toxical symptoms, but do not hesitate, if occasion requires, to push your medicines to the utmost limit which the patient will tolerate. I believe more harm is done than is generally known, in many cases, from the surgeon being afraid to use mercury in sufficient quantities to control the disease, and in syphilis, you must recollect, mercury, instead of acting as a depressant, seems to possess the properties of a tonic—indeed, it is the sheet-anchor in treatment. The same is true of the iodide of potassium, so far as regards its tonic property ; of little if any value in the earlier stages of syphilis, in the later (ulcerative) periods it is invaluable, but only as an adjuvant ; it never will take the place of mercury. Give it at the commencement in 10 grain doses, gradually increasing the amount until the symptoms are controlled or iodism occurs. This is characterized by coryza, lachrymation, and an eruption of papules and pustules on the face and shoulders (acne), and occasionally, though very rarely, by blebs. As to the amount, it may perhaps surprise you to hear how much of this salt patients with advanced syphilis will stand ; it is sometimes enormous. For example, in the case of Quinn, the patient I showed you with nervous syphilis, in whom the symptoms were distortion of the face and paralysis of the leg and arm of one side, attended with severe pain in the head and insomnia, the amount given was 120 grains at each dose, and this was repeated three times daily. In addition to this, he used a drachm of mercurial ointment every night by inunction, and perhaps you remember that when, after ten days of such treatment, I presented him to you again, the facial paralysis had almost entirely disappeared, the arm and leg had regained a great deal of their power, and he had lost much of the cachectic appearance which he formerly showed. And yet the case at first looked anything but promising, and only shows the importance in these advanced cases of large doses of the salt. Large as the above amount is, it is not as great as I have sometimes used, and I will formulate here some axioms which may be of use for you to remember in the treatment of syphilis :

Mercury is the main-stay in treatment, not only in the earlier but in the later stages as well.

Iodide of potassium is of little service in the earlier stages ; in the later stages, although of extreme value, it only assists in dispelling symptoms ; to produce radical effects, it should be combined with mercury.

In giving both mercury and iodide of potassium, watch your patient well to obviate the occurrence of toxical symptoms, and do not hesitate to use either remedy in sufficient amount to dispel the symptoms, no matter what the requisite dose may be.

You will oftentimes find in the graver forms of the disease, such as gummata or nerve-syphilis, that doses of 20 or 30 grains produce little effect ; carry your dose up to 50 or 60 grains, and you will have the gratification of seeing your patient improve at once.   What the mode of action is I cannot tell, for curiously enough, when given in large doses, nearly all the iodide of potassium given can be collected in the urine ; thus, if a dose of 60 grains be given, 40 of it will be excreted, leaving 20 to be absorbed, and yet if you give only 20 grains instead of 60, it makes no sort of impression on the disease.

It sometimes, though rarely, happens that the patient, through some idiosyncrasy, cannot tolerate iodide of potassium ; in those cases, the simple tincture of iodine may be used as a good substitute.   It should be given in the following prescription :

> ℞.   Tinct. iod......................... ℥ ss.
> Syr. fusci........................ ℥ iv.
> M.

Sig.—One teaspoonful, well diluted with water, three times daily, after eating.

This preparation is usually well borne by the stom‹ ach, and is by no means unpalatable.

The amount of this should also be increased precisely in the same way as the iodide of potassium, although the amount required will probably not be as large.

When we were discussing the natural history and symptoms of syphilis, I spoke to you about what is known as syphilitic cachexia, a condition characterized by lardaceous changes in the viscera. This is a very grave and serious disease,—serious, because the system refuses absolutely to absorb either food or medicine. When this occurs, the treatment by mercury and iodide of potassium has, if continued, to be combined with tonics and stimulants, which should be given with a liberal hand. Of the tonics, the principal ones are the ferruginous preparations, either alone or combined with cod-liver oil, and among the stimulants, the more diffusible ones, such as champagne and brandy ; but when a patient arrives at this stage of the disease there is little hope, and all that there is left for the surgeon to do is to make the road to the grave as easy as possible.

As regards the duration of treatment in the later stages, it must of necessity be prolonged, as the symptoms are more obstinate in character than in the early part of the disease. The patient should be prepared to continue his treatment for a year, and longer if occasion requires ; and this, too, even if all symptoms have disappeared, varying in this regard from the treatment given in the early periods of the

disease. After treatment by anti-syphilitic remedies has been continued as long as the surgeon deems
necessary, the patient should be subjected to a thorough course of tonics, in order to complete what the
mercury and iodide of potassium have begun.

Before closing this lecture, let me say a few words
to you in regard to prognosis. In the majority of
cases it is good ; patients recover entirely from their
disease, oftentimes without showing any of the serious lesions such as you meet with in the wards of
hospitals, and examples of which I have already
shown you. By recovery, I mean that patients show,
after the disease has run through a certain course,
no further symptoms of syphilis, even though they
have been kept under observation for several years ;
and should they marry, their offspring show no
sign or taint of disease so far as syphilis is concerned. It is not in the acquired form of syphilis
that fatal results occur so much as in the hereditary
form, where the mortality is large, and where even,
should the child survive to puberty, it is liable
throughout its whole life to show symptoms of its inherited malady.

In short, you may accept the following rules as a
tolerably good guide in cases of acquired syphilis :

*The average case of syphilis runs its course in from
eighteen to twenty-four months.*

*Under proper and careful treatment, the graver
forms of the disease seldom occur.*

After the disease has apparently run its course, and anti-syphilitic treatment has been suspended, the patients should be kept under occasional observation for another eighteen months, and if in that time no symptoms make their appearance, they may make their minds easy as to the future. This, you see, embraces a period of three and a half years, half of which is devoted to the disease, and the other half to watching for further developments.

These rules, you understand, are not absolute; indeed, none such can be given, but I believe they will serve as tolerably safe guides for you to follow.

6

LECTURE IX.

WE have heretofore considered only the various phases of acquired syphilis. To-day's lecture will be devoted to a consideration of the infantile and hereditary forms of the disease.

Hereditary syphilis may be divided into two principal groups : first, where it occurs at or shortly after birth ; and secondly, where it shows itself during childhood and adult life. There seems to be two notable periods of explosion—viz., at birth and at the period of puberty. We will commence with the first of these—syphilis at or shortly after birth.

When the disease shows itself at birth, the child may be either born dead, or, if alive, it usually succumbs in the course of a few days. The body is covered with large bullæ, filled with serum mixed with blood. These bullæ speedily break, evacuate their contents, and the epidermis covering them exfoliates, leaving a red, denuded surface beneath. This constitutes what is known as pemphigus neonatorum syphiliticum. When the disease in the mother is not very far advanced, the child may be born, to all appearances, sound and healthy, not developing any signs of the disease until some weeks or even months after birth. Of course, the longer the symptoms are delayed, the greater are the child's chances of viability.

Syphilis in the infant appears almost always within the first six months of extra-uterine life ; in the great majority of cases, within the first three. After a time the child loses its plump and well-nourished look, becomes thin and querulous, refuses the breast, and an eruption of the erythemato-papular variety appears upon the body, legs, and arms, particularly upon the soles of the feet and the palms of the hands. Conjoined with this are mucous patches of the mouth, throat, axillæ, and about the anus and genitals. The child is afflicted with " snuffles," a genuine coryza of the nasal mucous membrane, which renders respiration difficult. The healthy cry of the infant is exchanged for a hoarse, stridulous noise, due to the invasion of the larynx by the disease, and the child sinks rapidly from exhaustion and inanition, or from a direct poisoning by the syphilis. An autopsy reveals interstitial changes of the internal organs, especially in the liver and lungs, corresponding with the early stages of the gummous period. Not infrequently the child, before death, may be attacked by convulsions, due to inflammation of the meninges of the brain or spinal cord.

If the symptoms are not developed until at or near

the sixth month, they are less formidable in their course, being confined to manifestations corresponding with the earlier stages of acquired syphilis. These consist of the erythematous and papulo-pustular eruptions of the skin, conjoined with the moist secreting lesions of both mucous membrane and skin, which, from their delicacy in infants, are peculiarly liable to be attacked. Under vigorous treatment, the disease gradually subsides, and the child passes through the earlier years of its life with only occasional outbreaks, until the period of the second dentition arrives, when certain changes occur. Before that period, however, there are certain peculiarities of physiognomy which deserve attention. The forehead is very prominent and bulging ; the bones of the face appear abnormally small, those of the nose are sunken, and the child has a wizened and aged appearance ; the angles of the mouth are more or less deeply scarred, and the skin has an unhealthy, sallow look, different from the wholesome, clean complexion of sound children.

At the period of the *second dentition*, the *permanent* teeth (not the deciduous, remember) are apt to be defective and bad ; especially is this the case with the upper and lower central incisors. Sometimes they are notched at their cutting edge, and this may go so far as to present the appearance of a crescent ; at other times they stand widely apart, and the ends are bevelled off to quite a narrow edge, presenting what are known as "screw-driver" teeth. Both or only one of the incisors may be thus attacked. The lower incisors, instead of being even and sharp, are irregular, jagged, and serrated, like the teeth of a saw, while the other teeth, the bicuspids and molars, are frequently black, and crumble away to the edge of the gum.

Together with these diseases of the teeth, the eyes are liable to be attacked with both keratitis and iritis. Of the former there is one peculiar variety, which is nearly always associated with inherited syphilis, and is known as *interstitial punctate keratitis.* This form of disease begins in the interstitial layer of the cornea, rapidly invades Descemet's membrane, and appears as numerous minute white dots scattered throughout the tissue. Connected with it there may or may not be ulceration of the upper layers of the cornea. This form of disease is insidious in its attack, and is seldom attended with much inflammation of the conjunctiva.

Syphilitic iritis of hereditary origin is a serious matter, as it is usually attended with an abundant effusion of lymph, which may result in completely blocking up the pupil and rendering the patient blind. Even if it do not go as far as this, adhesions nearly always occur between the free edge of the pupil and the anterior capsule of the lens. The disease rapidly spreads to the deeper tunics of the eye, the choroid and the retina, producing serious impair- ment of vision.

As the child advances to puberty, the bones and nervous system begin to show the effects of syphilis, and the child will present enlargements of the tibia and ulna, or of the cranial bones, or else an extensive ulceration will occur in the soft palate and posterior wall of the pharynx. The nodes of the bones often break down and suppurate, and at the bottom of the ulceration thus formed, diseased bone will nearly always be found.

One form of bone lesion occurring in hereditary syphilis is of peculiar interest. I refer to the form known as *dactylitis syphilitica*, and which has been well described by Dr. Taylor, of New York, in a monograph published in 1875. This consists in an enlargement of the phalanges of the fingers and toes, generally the first, which may increase the bone to three or four times its original size. It is also attended with some degree of redness and pain. If left untreated, this swelling breaks down, opens in one or more places, and is frequently associated with dead bone. It is analogous to what happens in the late stages of acquired syphilis, and may indeed be regarded as a gumma of the periosteum of the bone. Under treatment the swelling subsides to a great degree, but in very few cases entirely, some thickening being left behind ; and where the joint is also affected, a stiff and deformed finger is but too often the result.

If the nervous system is attacked, the disease usually shows itself as epileptiform convulsions or chorea, and unless the surgeon be aware of the possibility of syphilis as the underlying cause, he is apt to regard it as a case of struma or scrofula.

The child, under proper treatment, may entirely recover from these symptoms, but, under any circumstances, will always be delicate in health, unable to withstand the attacks of intercurrent diseases, and liable to succumb to what would otherwise be trivial illnesses. In fact they are rotten, their inherited disease continually keeping them on the dividing-line between health and disease. Should they be fortunate enough to reach advanced age, they are liable to ulcerations of the bones and to nervous diseases.

With regard to the *etiology* of hereditary syphilis, there is, even at the present day, a vast difference of opinion. Many able writers contend that the father frequently is the sole cause of the disease in the child, without the mother becoming herself infected ; in other words, they claim that the semen of the father will infect the ovum without conveying the disease to the mother. I avow myself an absolute disbeliever in this doctrine ; I do not believe that the mother can give birth to a syphilitic child without being herself diseased ; and I hold that if the mother be not syphilitic, the children are not, no matter what the father may be. As in lectures like these it would be impossible to enter into a lengthy discussion of the *pros* and *cons* of the case,

I must be satisfied to tell you the bare fact, and to express the belief that your future practice will confirm my statement. Be it correct or not, here is the practical point for you to remember when you are called upon to treat syphilitic babies: include the mother in the treatment as well; the father also, if you can, but the mother always, else you will be chagrined to find that subsequent pregnancies are followed by syphilitic children.

Syphilis, in its earlier stages, especially if it be of a mild type, may show very few and slight symptoms, and even if the manifestations attract notice, the woman, from notions of delicacy, or more frequently from ignorance of their importance, will give the surgeon no history whatever. Remember, also, that the earlier lesions leave no traces behind them, and this, conjoined to the fact that pregnancy often exerts an influence in holding the earlier manifestations of syphilis in check, leaves the surgeon absolutely in the dark as to the cause of the child's syphilis. He then turns to the father, and if the unlucky wight has happened to have contracted syphilis as a bachelor, although before marriage he has entirely recovered, the disease of the child is laid upon his shoulders, to the great comfort of the surgeon and the edification of all concerned, except, perhaps, the father. Sometimes, however, he absolutely denies any previous disease, and the case is then consigned to the limbo of unknown causes.

Syphilis also is a fruitful cause of abortions ; and where, in any given case, repeated pregnancies have ended in miscarriage, it should always be regarded as suspicious, and the idea of syphilis being the cause, entertained, no matter if the woman at the time shows no symptom of the disease.

The treatment in these cases, to be of any avail, must be prompt and thorough ; and here, as in the acquired form of syphilis, mercury is the main reliance. It is of little use to attempt to treat the child through the mother's milk—that is, by putting the mother upon treatment ; because, in the first place, it is very doubtful if the mercury is excreted by the mammæ, and, in the next place, if it be, the amount is very small—too small indeed to be of any service. The inunction method in this form of the disease is by far the best, and should be practised in the following manner : a drachm of the oleate of mercury (20 per cent.) should be evenly spread upon a piece of cloth or thin flannel a foot wide, and long enough to go around the baby's body ; this should be applied like a swathe, and the mercury should be renewed every second or third day. Children in this condition stand mercury remarkably well, and the only care taken should be to see that this strength of the ointment does not irritate the skin ; if it does, a weaker solution of the oleate should be used, or else freshly prepared mercurial ointment. In addition, minute doses of either the bichloride of

6*

mercury, or of gray powder, may be given internally, —the bichloride in doses of from $\frac{1}{100}$ to $\frac{1}{50}$ of a grain, three or four times daily, in milk which has been diluted with one-half its own quantity of water, and slightly warmed ; the gray powder in 3 or 5-grain doses, placed directly upon the tongue. Iodide of potassium in these cases is useless, and the treatment should be confined to the use of mercury alone.

As regards the child's nursing, no one but the mother should be allowed to suckle it, inasmuch as the mucous patches which are nearly always found in this stage of infantile syphilis are eminently contagious, and you have no right to expose an otherwise healthy woman to the risk of infection. It is a curious fact, which was pointed out as early as 1837 by Dr. Abram Colles, of Dublin, that the mothers of syphilitic children, although they themselves may show no signs of the disease, are not obnoxious to contagion from syphilis ; hence, the child may, with impunity, suckle its *apparently* healthy mother, where it would be a source of danger to any healthy stranger who should attempt to perform the maternal function. If the mother should be unable to suckle the child, it must be weaned and brought up on the bottle.

Supposing the child to recover from its earlier symptoms—the mercurial treatment having been continued, of course, until all manifestations have disappeared—it should be placed upon a tonic treatment, and kept under observation for a couple of years. It may then be dismissed, with the injunction to the parents that fresh symptoms must be expected when the child arrives at the age of puberty; and should any manifest themselves either at that time or before, the child must at once be placed under medical observation. As I have already pointed out to you, the symptoms which present themselves at the period of puberty are analogous to those which occur in the later stages of acquired syphilis—viz., nodes of the bones, diseases of the nervous system, and ulcerations of the mouth and throat. Here it is that you find the *iodide of potassium* coming into play as a feature in the treatment, but *not*, I beg you to remember, to the *exclusion of mercury;* for you will obtain the best results where you combine the two.

This may be done in the following manner :

℞. Hydrarg. bichlor.............. gr. ss.–i.
Kali iodidi................. ℥ i.
Syrup. aurant. cort., or syrup. sars.
comp.......................·q. s. ad ℥ ij.
M.
Sig.—Teaspoonful in water, three or four times daily.

If it be preferred to give the two separately, the pill of the protiodide of mercury may be given once daily in one-quarter or one-third of a grain dose,

and the iodide of potassium twice or thrice daily, after
meals, in the following prescription :

R̲.   Kali iodidi.................... ℥ i.– ℥ ij.
    Syrup. sarsæ comp............ ℥ ij.
M.
Sig.—In teaspoonful doses.   .

If the iodide should not be well borne, the tincture
of iodine may be substituted as follows :

R̲.   Tinct. iod..................... ℥ ij.–iv.
    Syrupi fusci.................. ℥ iij.
M.
Sig.—Teaspoonful three times daily.

All these preparations in which the syrups enter
should be made in small amounts, and freshly prepared, as the syrup is liable to undergo fermentation
when long kept.

In the intervals of the mercurial and iodic treatment, one of the best tonics for internal administration is the syrup of the iodide of iron, either alone
or in combination with cod-liver oil, and the syrup
of the hypophosphates of lime or soda.   This latter
is particularly to be commended in the osseous and
nervous lesions of inherited syphilis.

The following prescriptions will be of service :

R̲.   Syr. iod. ferri..................... ℥ ij.
Sig.—Five to ten minims three times daily, after meals.

℞.   Syr. iod. ferri...................... ℨ iv.
      Ol. morrhuæ....................... ℥ iij.
M.
Shake well before using,
Sig.—Teaspoonful after meals, thrice daily.

℞.   Syr. calcis et sodæ hypophosph...... ℥ ij.
In teaspoonful doses twice, or thrice daily.

This may be combined with cod-liver oil, if deemed desirable, in the same doses as given above.

The old manner of giving the iodide of iron in pill form—what is sold under the name of Blancard's pill—is not as good as the syrup, inasmuch as the pills, if kept for any length of time, are apt to get hard, and are not easily acted upon by the intestinal juices.

LECTURE X.

GONORRHŒA OF BOTH SEXES.

GONORRHŒA, or, as it is more commonly called, clap, is one of the most common forms of venereal disease which you will be called upon to treat, and oftentimes one of the most obstinate and rebellious to treatment. I shall consider it, first, as it affects the male ; secondly, as it affects the female.

Gonorrhœa in the male is a muco-purulent discharge from the urethra, generally due to irritation caused by a similar discharge in the female. This is one of the commonest causes of the disease in the male ; but others have been assigned, and first among these is leucorrhœa. This, in its acute form, is often very difficult to distinguish from a gonorrhœa, inasmuch as they are both attended with extreme inflammation and redness of the vaginal membrane, as well as with a profuse purulent yellow discharge. In addition to these, the menstrual flow is regarded as capable of inducing a urethral discharge in the male ; and, paradoxical as it may sound, a man may contract an inflammation and discharge in these parts from a perfectly healthy woman. I admit that such cases are very far from common, the usual source of the disease being, as I have stated above, from a co-existent clap in the female.

Gonorrhœa occurs without any period of incuba-tion, usually appearing within forty-eight hours after the suspicious coitus ; and the first noticeable symp-tom is a slight tickling just within the meatus, which becomes more marked during micturition. If pres-sure be exercised along the floor of the urethra, a drop or two of sticky fluid can be squeezed from the end of the penis. This matter is thin, colorless, and does not stain the linen. After twenty-four to forty-eight hours have elapsed, the discharge will be seen to lose the characteristics just detailed, to be-come thicker and white like milk ; and the act of mictu-rition is more painful. If the disease be left to itself, the discharge becomes more and more abundant, sometimes so much so as to drip from the patient; it loses its white appearance and becomes yellow, and, if the inflammation is acute, of a greenish or rusty hue, from the admixture of blood. The act of uri-nation now becomes decidedly painful, the stream of water very much diminished in size, and when the inflammation is high the water is only passed drop by drop. Where this condition of things obtains, febrile symptoms are often present, particularly in a first attack, attended with a high pulse, hot and dry skin, and a furred tongue. The penis is œdematous and swollen, and where the prepuce is long the œdema may be so great as even to cause partial or complete phimosis. The lymphatics on the dorsum penis are enlarged, red, and painful, and the glands in the groin may also participate in the general inflammation — becoming, in their turn, swollen and tender.

In other cases, the inflammatory symptoms may be entirely absent, the only signs present being painful micturition and a purulent discharge. This usually reaches its height about the tenth day, invading in its progress the urethral mucous membrane from the fossa navicularis, the starting-point of the disease, to the region of the bulbus urethræ. At this date the discharge retains its yellow character, but the act of micturition is less painful than during the first few days of the clap. It then remains stationary for another ten days or so, when the discharge gradually loses its purulent and yellow character, changing to white, and from that to a thin, viscid, colorless flow, running, in other words, but in a descending scale, through the same course that it pursued in its commencement. As the discharge becomes less and less purulent, the act of micturition becomes easier, until finally all pain and discomfort cease during the act. Gradually this thin discharge diminishes, until it finally dries up entirely, and the patient finds himself well. This is the course usually pursued where no complications are present; where these occur, however, the duration of the disease is much more prolonged, more painful and serious, but of these I shall speak in a separate lecture.

In the female, the disease shows itself first as an inflammation of the vulvar mucous membrane, invading the vestibule and the labia majora et minora. Micturition is attended with some smarting and pain, due to the acid urine passing over the irritated and inflamed mucous membrane, and not to any disease of the urethra itself. Where you find the urethra in women the seat of a muco-purulent discharge, you may say with confidence that the disease is gonorrhœa, for no leucorrhœa that I am conversant with is attended with a discharge from the urethra. Attendant upon this inflammation of the mucous membrane of the vulva, is a thin, viscid, colorless discharge, analogous to what occurs in the male, which speedily becomes thick, abundant, and purulent, staining the woman's body-linen. This inflammation rapidly extends from the vulva along the vagina, which upon examination is seen to be red, swollen, and secreting an abundant amount of thick, yellowish pus. The temperature of the parts is also increased. This inflammation may extend to the mucous membrane lining the cervix uteri, and even to the uterus itself, producing serious symptoms. After lasting for several weeks, the discharge diminishes in intensity and purulence, and the mucous membrane of the vagina becomes less red and swollen, although the discharge may continue for some time longer. It now, however, becomes of a light yellow or white color, closely resembling an ordinary leucorrhœal discharge. When it arrives at this stage, unless subjected to treatment, it remains stationary for a long time, being liable to exacerbation from various causes, until it gradually wears away to the thin, viscid discharge which marked the advent of the disease. Pain is no longer felt during micturition, as the mucous membrane of the vulva becomes thickened, and is no longer sensitive.

As regards the etiology of this disease, I wish to explain to you a little more fully what I told you in the beginning of this lecture. It is easy to understand how a clap in a woman may produce the same in the male; but why the menstrual flow, or why a perfectly healthy woman should be capable of exciting trouble in the male, is not quite so easy to comprehend. In the first place, you must distinctly bear in mind that gonorrhœa is not produced by any virus, such as we understand the term when speaking of chancroid and syphilis; it is a pure and simple catarrhal inflammation, and may be produced artificially in a healthy person, by the use of irritating injections, and even by the improper use of sounds. If now a man has connection with a woman during the menstrual period, more particularly at its commencement, or toward its close, this flow, from its irritating nature, may sometimes produce an inflammation of the urethral mucous membrane in the male. I confess to some degree of scepticism in cases alleged to be produced from this cause, but as it is reckoned among the possibilities, I give it to you here for what it is worth.

The causes which produce clap from connection with a healthy woman are due much more to the man than to the woman. In your future practice, you will often be told some story like the following : The man, after dining with a party of friends, and having drunk freely at dinner, has intercourse with a woman of the town. Afraid of the consequences, and desirous of averting them as far as possible, he borrows a favorite clap prescription from some friend, and injects it industriously into his urethra, when, after the lapse of a few days, he is much disgusted to find the very disease appearing which he fondly hoped he had averted ; in other words, he has given himself a clap from his officious over-medication. The woman is examined, and the parts are found healthy, nor is there any reason to believe that she was the cause of the trouble in her companion. To speak strictly, the woman has had nothing at all to do with inducing the disease ; it has been entirely the man's fault.

Besides these causes, Ricord has stated that sexual excitement, without contact or any other kind of irritation, might produce a urethral discharge in the male, and he gives one case which stands unique in venereal literature. It occurred in the person of a physician, who, from ten o'clock in the morning until seven in the evening, vainly endeavored to overcome the virtuous scruples of a young woman with whom he was in love, and during all this time he was in a condition of extreme sexual excitement. Three days afterward he was attacked with a most painful and violent clap, which lasted for forty days. Let me add that previous to this unlucky tête-à-tête, it is stated that the physician had been continent for six weeks. I candidly admit that I do not believe the story. I think that Ricord was willing to be charitable to a brother physician ; and had it occurred in the person of any but a confrère, he would have made it the subject of ridicule. At any rate, no similar cases have ever been reported to my knowledge.

Let me tabulate the following axioms for you :

Gonorrhœa does not depend upon a virus ; it is a simple catarrh of the urethral mucous membrane, and is due to the presence, within the canal, of some local irritant. Gonorrhœa is produced by gonorrhœal, leucorrhœal and the menstrual discharges.

A perfectly healthy woman is reputed to be capable of producing urethral inflammation in the male ; such cases, however, should not be admitted without some reserve.

Gonorrhœa has no period of incubation.

You would be wrong should you consider that every urethral discharge in the male is necessarily a clap ; undoubtedly the majority of such diseases are gonorrhœal, but you may also have other causes at work to produce this condition of things. A urethral chancroid or initial lesion will produce a running from the genitals, and it is oftentimes difficult to decide at once whether the disease under observation is really a simple clap or not. If the cause be due to a concealed chancroid the following symptoms will serve to put you upon the right track. The pain in the urethra is localized, and not general as it is in clap ; the discharge, although purulent, is not very abundant, and is frequently streaked with fresh blood, and pressure along the floor of the urethra excites pain only at the seat of the lesion. The crucial test, however, is auto-inoculation. If the matter be due to chancroid it is capable, by inoculation, of producing another chancroid, while gonorrhœal pus is innocuous. Separation of the lips of the urethra will often discover the sore seated just within the meatus ; but if, as sometimes happens, it is situated deeper within the canal, the examination must be conducted in another manner. An instrument called the meatoscope should be passed a short distance into the urethra as far as the fossa navicularis, and a strong light thrown into the canal with a reflecting mirror, when the lesion will be brought clearly and plainly into view. An ordinary ear speculum, by the way, makes one of the best instruments for examination— better, indeed, than nine-tenths of the meatoscopes offered for sale in the shops.

If, however, the discharge be due to a concealed initial lesion, the symptoms are somewhat different; the discharge is very thin, and seldom becomes purulent, unless irritated from some cause or another. Palpation reveals, in the great majority of cases, an indurated spot in the course of the canal, and an examination of the urethra in the method already advised gives the clew to the proper source of the urethral discharge.

Besides these causes, gouty persons are very liable to slight discharges from the urethra; and especially is this the case after the patients have indulged a little more freely than usual in the pleasures of the table, particularly in the use of heavy-bodied wines, such as Burgundy or port. Here the disease comes on without any history of sexual indulgence, and is attended with pain during micturition near the neck of the bladder and along the course of the prostatic urethra. The discharge which accompanies this form of disease is not very abundant, although it is slightly purulent; it stains the patient's linen, and comes from the deeper part of the canal—never from the anterior portion, as in clap. The urine is very acid, and loaded with urates. Under proper treatment, these symptoms abate in the course of a week, and leave the patient as well as he was before.

Tight strictures of the deep urethra may also cause a muco-purulent discharge; but as a consideration of these diseases belongs rather to the domain of surgery than of venereal medicine, I shall content my-self with a mere mention of them as an exciting cause.

The duration of a clap varies very much. As I have already told you, in the female it may last for several weeks, and even months, and is one of the most obstinate diseases to treat, owing to its liability to invade the mucous membrane of the cervix uteri, where it is difficult to reach it by local remedies, and internal treatment has but little if any efficacy in vaginal and cervical gonorrhœa.

In the male, however, although sufficiently obsti-nate, it is not so chronic as in the female ; and this is due, in a great measure, to the greater care and per-sistency with which the treatment is followed up. In man the disease, unless complicated, usually runs its course in from four to six weeks ; but if any of the complications supervene of which I shall speak to you in the next lecture, the disease may be prolonged for eight, ten, or more weeks. Much depends upon the attention and fidelity with which the patient carries out the treatment, and he should be particularly cau-tioned to continue it for a short time after apparent recovery has taken place, because a clap is very prone to relapse, and each relapse makes the disease more difficult to cure.

LECTURE XI.

THE complications which occur in gonorrhœa are numerous, and some of them quite serious in their nature. The first one which I shall consider is *balanitis*, or an inflammation of the mucous membrane of the prepuce and the glans penis, which is characterized by intense redness of these parts, and is attended with superficial excoriations, which may be easily mistaken for superficial chancroids, or for mucous patches; but the inability to inoculate their secretion, their superficial character, and the facility with which they get well under the simplest treatment, will prevent their being mistaken for the former; and the absence of concomitant symptoms, as well as all history of syphilis, will exclude them from the category of syphilitic manifestations. They usually appear as mere erosions of mucous membrane, and seldom, unless irritated, are they covered over with any secretion. If a pellicle form over the abraded points, it can usually be readily removed.

Phimosis and *para-phimosis* may occur in those cases where the inflammation is very acute, and sometimes go so far as to produce serious inconvenience and even danger. In phimosis, the prepuce is incapable of retraction, and the discharge from the urethra as well as the urine collects within the constricted foreskin, requiring the utmost attention on the part of the patient to cleanliness, in order to obviate ulceration and sloughing of the prepuce.

Where the prepuce is naturally short, if the parts become inflamed and swollen the constriction takes place behind the glans penis at the fossa glandis, and, unless relieved, may go on to gangrene and sloughing of the glands. This portion of the penis becomes purple, the temperature is diminished, and the parts slough from the mechanical obstruction to the circulation.

Sometimes along the course of the urethra one or more points may become hard and exquisitely tender, which, after a time, soften and break down, discharging a quantity of laudable pus. These are known as peri-urethral abscesses, and are usually found at the frenum, the peno-scrotal angle, and the perineum, although they sometimes occur at intermediate points. They are usually ushered in with a chill and a slight rise in temperature, which is speedily followed by the presence of pus in the swelling. Attaining to large size, they often press upon the urethra in such a way as to diminish its calibre and interfere seriously with the act of micturition.

One of the most frequent complications of gonor-

7

rhœa in the male is known as *chordee*, which is a painful curvature of the penis during erection. This may take place in three ways : with the concavity looking downwards, upwards, or sideways, and is due to. an exudation of lymph into the corpus spongiosum or the corpora cavernosa. This distressing symptom comes on only during erection, and seems to be particularly favored by the heat and warmth of the bed. Sometimes the amount of inflammation and distortion which occurs is so great as to produce free hemorrhage from the urethra, leading to temporary relief, but as soon as the local effect has passed off, the chordee returns as vigorously as ever.

After a clap has lasted for three or four weeks, invading the deeper portion of the urethra, the patient begins to complain of uneasiness and pain in the testicles, and, upon examination, these organs are found to be enlarged and tender. Although the name of orchitis has been given to this affection, the body of the testicle itself is not implicated, but only the epididymis, which in this stage of the disease is attended by the usual symptoms of pain, redness, and swelling. You remember, when we were discussing syphilis, I mentioned a form of epididymitis which occurs in that disease, and I wish to call your attention to the diagnostic points of difference which obtain between the two varieties. In syphilitic epididymitis this body is indurated, but is devoid of pain or redness, indeed, so little uneasiness is there, that the part can

be freely handled without inconvenience to the patient, but in the gonorrhœal variety the epididymis is
red, swollen, and exquisitely tender, so much so that
the mere contact of the bed-clothes is sufficient to
excite pain and discomfort, and I need hardly add
that free handling of the part is impossible. This
generally comes on about the third or fourth week
of the duration of the clap, and during its continuance the urethral discharge almost entirely disap-.
pears—to reappear, however, upon its subsidence.

The acute inflammation lasts from seven to ten
days, at the expiration of which time it gradually
subsides, leaving the epididymis indurated, although
not very sensitive, and this induration may be further
complicated by the effusion of fluid between the two
layers of the tunica vaginalis, constituting what is
known as *hydrocele.* Under proper treatment the
fluid is absorbed, and the swelling of the epididymis
diminishes ; indeed, under very favorable circumstances, it may entirely disappear ; but this result is
not always attained. Only too often the epididymis, as well as the vas deferens, is permanently
blocked up, preventing the egress of the spermatozoa from the affected testis, and leading to partial
sterility. Instead of resolution, one other course may
be pursued : the part may suppurate ; and when it
does, destruction of the epididymis, and sometimes
of the testicle on that side, follows.

The disease is usually unilateral, one testis being

affected pretty nearly as often as the other ; but sometimes it is double, when, of course, it becomes more
serious, inasmuch as the induration and obliteration
of the canals of the vasa deferentia lead to permanent
sterility. I beg you will distinctly understand the difference between sterility and impotence ; the sterile
patient is not rendered impotent, he is capable of per-
fect connection even to the emission, but the semen
ejected is devoid of spermatozoa—in other words, he
is incapable of procreation ; while the impotent man
is incapable of connection, although his semen is
fruitful. When one testis only is affected, the patient can still be the father of children, but only as
regards his sound testicle.

The inflammation may extend from the testis to
the spermatic cord, and when this is the case the
patient complains of pain running from the testis to
the lumbar region, with a dragging sensation upon
the cord, as though traction were being exercised
upon it. An examination reveals a thickened condition of this portion of the genital apparatus, which is
sometimes enlarged to the size of a goose-quill, and
excessive tenderness, with inflammatory redness running up as far as the ring. Treatment usually causes
these acute symptoms to abate in from five to ten
days ; the thickening, however, lasts longer, until
finally it entirely disappears, and the cord resumes
its normal condition.

In rare instances resolution does not take place ;

but instead of this, suppuration occurs somewhere along the course of the cord external to the ring. When this takes place, there is danger of atrophy of the testis, resulting from obliteration of the spermatic vessels.

Later in the disease other portions of the genito-urinary apparatus may be affected, and the prostate is the next organ to feel its effects. This body is, as you know, composed of both muscular and glandular tissue, and encircles the neck of the bladder ; and from the intimate connection between the two, one rarely escapes when the other is attacked. Hence, I shall consider the two together, although they are often treated in works on Venereal Diseases under the separate heads of *prostatitis* and *cystitis*. The first symptom which the patient notices is a sensation of uneasiness rather than of actual pain in the perineum, together with a feeling of weight and tension in the part, and this is particularly noticeable when he sits down. This symptom gradually increases until both the erect and sitting posture is painful, and the patient only finds relief when lying upon his back. Connected with this is a still more unpleasant symptom, viz., a constant and urgent desire to pass water, which comes upon the patient so suddenly and violently that, no matter where he is, he has to respond at once to this call of nature, and the urine is voided quite as frequently in his clothing as out of it. After the water is ejected—and nearly always this occurs in very small quantities—there is a violent straining and bearing down, which is present not only at the neck of the bladder, but in the rectum as well, as though the bladder and rectum needed instant evacuation. This is known as tenesmus, and may be so violent as to cause hæmorrhoids or a prolapse of the bowel.

Upon examination through the rectum, the prostate will be found enormously enlarged, encroaching upon the bowel, and most exquisitely tender, and this inflammation may pursue one of two courses: either the symptoms entirely subside and the disease passes off, or an abscess of the prostate may result. If this occur, it terminates either by breaking into the urethra—the most favorable of all courses—or else it opens into the rectum, producing a recto-prostatic abscess; or if, as sometimes happens, the abscess opens in both directions, a fistula between the urethra and rectum is established, which is extremely difficult to cure.

Another complication is what is known as *cowperitis*; an inflammation of two little glands seated anterior to the prostate, the ducts of which open into the urethra. Sometimes this occurs alone, but it is more frequently associated with prostatitis. The symptoms in the two diseases are much the same, and the course they pursue is very similar.

The vesiculæ seminales are also liable to be attacked, when the patient will complain of deep-seated pain in the perineum, usually upon one side. An examination per rectum, carrying the finger well up alongside of the bladder, reveals a tender and swollen spot, which after a time gradually subsides, or else it ends in suppuration.

Toward the end of a clap, when the discharge has become thin and colorless, when the anterior portion has entirely recovered its normal condition while the posterior portion of the urethra still remains diseased, a condition of affairs arises to which the name of *gleet* has been given. Here the discharge, instead of being continuous as it is during a clap, is only seen on rising in the morning as a single drop of white or colorless matter, which does not stain the linen, and which is not accompanied with any pain during micturition. This drop of fluid is usually obtained only upon deep pressure, and during the day is absent. If the patient commits any excess in eating or drinking, or if he indulge in immoderate coitus, this drop may increase to a slightly purulent discharge, which lasts for a few days, and then subsides to its former condition. Examination with a bulbous bougie will reveal in the membranous or prostatic portions of the urethra one or more localized points of tenderness, which offer a slight resistance to the passage of the bougie, and which usually bleed. This is due either to a granular and thickened condition of the urethra, the incipient stage of stricture, or else to a slight stricture which has already formed. Remember, that nine cases in ten of gleet are dependent upon stricture; hence, when called upon to treat a gleet, always search at once for stricture, and generally the removal of that means the cure of the gleet. If no stricture be found, then the discharge is due to inflammation of the deep urethra, attended perhaps by slight erosions of the mucous membrane, which will require different treatment to what it would were it due to a stricture.

I have´ purposely omitted speaking at length on stricture, as this form of disease belongs more properly to the domain of genito-urinary surgery than to venereal medicine; and I only mention it here to show you how it may act as one of the underlying causes of gleet.

In the female the complications which occur in the course of a clap are not so numerous, but at the same time some of them are much more serious. One of the most frequent is that which occurs in the earlier stage of the disease known as *vulvitis*, and is analogous to balanitis in the male. It is attended with erosions of the mucous membrane of the vulva, the vestibule, and the fourchette, with a copious muco-purulent discharge, which, flowing over the perineum and the inside of the thighs, irritates and excoriates these parts. If the inflammation is very acute, the labia majora become œdematous and swollen, and sometimes it ends in suppuration. In addition to this the gland, which is known as the vulvo-vaginal, or gland of Bartholin, the duct of which opens just within the introitus vaginæ, becomes enlarged and painful. This may be felt as an ovoid swelling seated at the posterior commissure of the labia, is usually unilateral and nearly always ends in suppuration, following one of two courses, the pus either being evacuated through the duct into the vagina, or, if occlusion of the duct ensue, requiring an incision externally for its cure.

Although gonorrhœa in women is generally confined to the vulva and vagina, it sometimes happens that the urethra, as in the male, is the seat of trouble, either concomitant with the vaginitis, or else, very rarely, alone. Owing to this canal being shorter in the female than in the male, urethritis in women produces but little disturbance beyond an urgent desire for frequent urination, and is seldom followed by cystitis. To detect its presence, the vagina being first thoroughly cleansed of all discharge, the surgeon passes his finger into this canal and makes firm pressure along the urethra from the neck of the bladder downward and outward toward the meatus, when a few drops of pus follow the finger. When you find this condition of things you may be sure you have to deal with a gonorrhœa (of course supposing there is no concealed urethral chancroid or initial lesion), for no other disease that I am aware of produces similar symptoms. If there be a concealed urethral ulceration, the symptoms will be the same as those detailed a

7*

few pages back, when speaking of similar lesions in the male.

You may, therefore, formulate the following rule:
Urethritis in the female is always due to some venereal affection.

I have already sketched the course which a gonorrhœal vaginitis pursues in an ordinary case, and have shown how it spreads from the anterior portions to the deeper parts, ending in an inflammation and purulent discharge from the cervix. But it may extend even beyond this region, and, invading the body of the uterus itself, extend along the Fallopian tubes and attack the ovaries. This inflammation of the uterus, or, as it is called, *metritis*, is always a serious matter, and commences with a feeling of congestion of the organ, attended by a severe bearing-down pain and disturbance of menstruation, which is usually scanty and difficult. If a recto-vaginal examination be made, the organ will be found swollen and exquisitely tender, and pressure above the pubes excites similar discomfort. High fever, increase in temperature, and a full pulse are nearly always present. As the disease passes along the Fallopian tubes symptoms of peritonitis present themselves, which may increase in intensity and result fatally. This termination is not a common one, although a few cases have been reported ; it usually ends in recovery.

When the ovaries are attacked pain is experienced over the region of one or the other of these bodies, which is intensified upon deep pressure externally or upon a vaginal examination. The skin covering the ovarian region shows signs of inflammation, and this form of disease is frequently attended with febrile manifestations, which after lasting a week or ten days, gradually subside and the parts resume their former condition.

Besides these symptoms, if the inflammation be very acute, the lymphatics and the inguinal glands in both sexes are implicated. In the male, the lymphatics running along the dorsum of the penis may be felt as a hard line, the size of a large goose-quill, running from the fossa glandis to the crura penis, and are there lost in the inguinal chain of glands. Their course may often be followed by the eye, appearing as a broad, red line overlying the inflamed lymphatics. After a while the inflammation subsides and the lymphatics are no longer apparent to the eye or to the finger, or else suppuration occurs at one or more points along their track, which, upon evacuation of the pus, usually heal readily.

This inflammation may extend to the inguinal glands when we have in one or both groins a tense, brawny, and inflamed swelling, painful to the touch, and a hinderance to locomotion. This enlargement may be ushered in with a chill and a slight elevation in the temperature. Sometimes the inflammation subsides without ending in suppuration, while again pus forms within the body of the gland and an abscess is the result.

Two other forms of disease resulting from gonor-rhœa yet remain to be spoken of—to-wit, rheuma-tism and ophthalmia. The occurrence of these affec-tions at a distance from the original lesion led to the belief at one time that gonorrhœa was a virulent disease, and that the virus, by absorption, produced these remote symptoms. It is now generally ad-mitted that this view is erroneous ; that these fibrous tissues are liable to attack just as is the fibrous tissue in the urethra of the male ; and this theory derives additional support from the fact that women who are very rarely indeed attacked by urethral gonorrhœa, seldom suffer from rheumatism.

Gonorrhœal rheumatism generally comes towards the end of a clap, although there are exceptions to this rule, and invades joints in preference to other parts—usually the knee, the elbow, and the wrist. Occasionally very acute, and attended with a marked degree of inflammation, its general course is a sub-acute one, with swelling and pain. Shortly after the access of the disease, effusion of fluid takes place into the joint, accompanied with an increase of pain, continuing in this condition for a time varying from two weeks to several months ; the fluid is gradually absorbed, and the joint may be restored to its former usefulness. Unfortunately, however, this is not al-ways the case ; ligamentous adhesion takes place, and anchylosis, partial or complete, is the final re-sult. I know of no cases more disheartening and annoying in the entire range of venereal affections than those of gonorrhœal rheumatism, both on account of their chronic course and because the results of treatment are but too often unsatisfactory.

The tendons come next to the joints in point of frequency of attack, and then the muscles. Of the former, the tendo-Achillis is the one most likely to suffer, and when it is attacked, the disease runs a long and painful course, not so much from swelling, for this is often trifling, but from the steady aching of the part and the consequent impediment to walking. In very severe cases a permanent contraction of this tendon results, producing a talipes equinus, for which tenotomy is the only relief.

Another peculiar symptom occasionally met with in this stage of gonorrhœa is a persistent, boring pain in the os calcis, unattended by redness or any enlargement of the bone or thickening of the periosteum.

It is chronic in its course, and is very apt to occur in nervous men. I recall one case where it had lasted for several years. Indeed I am inclined to regard it as a neurosis rather than a periostitis.

Until within a few years it was believed that *the heart* escaped in gonorrhœal rheumatism, differing in this respect from the ordinary form of rheumatism. This is, however, a fallacy ; the pericardial sac, as well as the valves of the heart, are affected. The patient complains of præcordial pain, attended

sometimes with dyspnœa, when an examination reveals an effusion into the pericardium, and the heart-
sounds are muffled. As this subsides, a souffle is
heard at the aortic and mitral valves, sometimes
with regurgitation.

Gonorrhœal ophthalmia has been divided in many
treatises on Venereal into two varieties : one which
is due to the presence of gonorrhœal pus in the eye ;
the other, analogous to what take place in gout, an
irido-scleritis rather than a true ophthalmia. The
latter is the only *bona fide* disease which belongs to
gonorrhœa, the first one, although by far the more
serious of the two, being nothing more than a purulent ophthalmia, due to an accidental infection.

The first symptom of which the patient complains
is a sensation of weakness in the eye ; this is very
seldom associated with photophobia, although occasionally this also may be present. Upon examination, the conjunctival and sclerotic vessels will be
found somewhat congested, the iris slightly infiltrated, with a sluggish pupil, the anterior chamber distended with fluid containing occasionally some floc-
culi. The tension of the eyeball is also increased.
As the disease progresses, the anterior capsule of
the lens, as well as Descemet's membrane, becomes
opaque and the cornea loses its transparent look.

This condition lasts for some days, when, under
proper treatment, the symptoms subside ; the iris,
the capsule of the lens, and the cornea, resume their

normal appearance, and the disease passes off, leaving the eye none the worse for the attack.

Not so, however, with the purulent variety. Here the situation is very grave, and, unless active measures are speedily adopted, the eye is irretrievably injured, the contents of the ball being evacuated in forty-eight hours, or even in less time. This disease is due to conveyal of the pus from the genitals to the eye, and the right one is the one most frequently affected, for the simple reason that there are more right- than left-handed people. The symptoms noticed are lachrymation, photophobia, intense congestion of the conjunctival vessels, together with a thick, purulent discharge. Both lids speedily become œdematous and enormously swollen, so much so as to completely close the eye. If the lids be gently separated, the conjunctival and palpebral mucous membrane will be found swollen and perfectly scarlet in hue. The former, from the swelling, is very much elevated above the cornea, leaving this latter imbedded in the inflamed tissue, like a watch-glass in its setting. This swelling is known as *chemosis.*

The cornea, curiously enough, is at first unaffected, but it rapidly, from pressure and interference with its nutrition, becomes opaque, pus forms in the interstitial layer, which, pushing through the epithelial covering, leaves behind ulcerations of the cornea ; this tissue softens, and the tension of the eyeball being great, the lens and vitreous humor are evacuated through the opening. In other words, the eye is completely lost.

While this is going on, the abundant purulent secretion is poured out over the cheeks, producing excoriation of the skin of these parts. Occasionally the pressure upon the lids is so great from the œdema that gangrene ensues, sloughing of the lids occurs, and greater or less deformity follows.

Under prompt treatment, thoroughly carried out, the eye may be saved; but opacity, with some ulceration of the cornea, nearly always results. The œdema subsides, the chemosis disappears, and the conjunctival congestion abates in intensity. A thickened and granular condition of the palpebral mucous membrane remains, however, for a long time after, which requires steady and constant care to cure.

Gonorrhœal discharges from the nose and mouth have been spoken of as occurring in some few rare instances. These are rare indeed, if they ever do occur, and the reported cases are by no means convincing.

In addition to the above-mentioned complications which occur in the course of a clap, there are two diseases which, although not strictly complications, are frequently found with gonorrhœa, or else are the indirect results. They are commonly known as *venereal warts* and *herpes*. The term *venereal* warts is another one of those misnomers which abound in the literature of venereal diseases; for although some-

times found with a gonorrhœa, they may be absolutely and entirely independent. They are usually
seated, in the male, upon the mucous membrane of
the glans penis, the inner lamella of the prepuce,
upon the scrotum, and sometimes upon the perineum
and the pourtour of the anus ; in the female, they
occur upon the mucous membrane of the labia majora
et minora, upon the perineum, and about the anus.
They occur as papillary excrescences, raised above
the surface of the mucous membrane, exceedingly
vascular, bright red in color, and, when favored by
heat and moisture, are of exuberant growth. They
are, indeed, nothing but hypertrophy of the natural
papillæ of the parts, and are particularly prone to
attack those who are careless of their personal cleanliness. They may attain to enormous size, and I have
seen cases where the head of the penis was transformed into an enormous bulbous mass, resembling
a cauliflower, entirely obliterating all semblance to the
ordinary virile member. Their shape varies somewhat
with their location, and when they are compressed,
as, for example, when seated on the perineum, or in
the cleft of the nates, they grow in the shape of a
cock's comb, being long, pointed, and serrated. In
the female, we find them most exuberant, and they
sometimes extend from the anus over the perineum
and vulva, up even into the groins, assuming the most
grotesque appearances, and from attrition and dirt
give rise to a very offensive and acrid discharge.

Herpes is another manifestation, although not strictly venereal in its origin, which it behooves you to know something about, inasmuch as it is frequently confounded with superficial chancroids or mucous patches· of the glans penis. It appears upon the mucous membrane of the prepuce and glans penis as a group of minute vesicles, five or six in number, seated upon a slightly inflamed base. These vesicles rapidly coalesce, and in the course of twenty-four or thirty-six hours are denuded of their epithelium, when they present superficial erosions, which are sometimes covered with a whitish pellicle. If seen early in their course, before the vesicles are broken, there will be no difficulty in recognizing the disease ; but when the vesicles have become eroded it is sometimes extremely difficult to distinguish them from superficial chancroids and mucous patches. Its non-auto-inoculability, the rapidity with which it recovers under simple treatment, its non-tendency to spread, and its history, will serve, in most cases, to prevent you from mistaking it for the first ; and the absence of all syphilitic history and concomitant symptoms of the pox, will save you from mistaking it for the second class of these diseases. It is sometimes due to local causes of irritation, but quite frequently it is associated with digestive disturbances, induced by over-indulgence in eating and drinking.

LECTURE XII.

OF all the venereal diseases which demand treatment at the hands of the surgeon, gonorrhœa is the
most uncertain and disagreeable. The number of
nostrums that have been sold for its cure are innumerable, and almost every medical man has some
pet plan of procedure, which, in the long run, turns
out to be neither better nor worse than others which
have been in use since gonorrhœa was recognized as a
curable disease. Some writers decry the use of injections as provocative of stricture, and undoubtedly,
if improperly used, they may do more harm than
good; while others assure their readers that injections are the safest and best remedies for the cure
of the disease. To this latter view I give my adherence; and, while recognizing the fact that local
remedies are the mainstay in the treatment of the
disease, I do not go so far as to exclude internal
treatment; not arguing from that, that the internal
treatment is essential to the cure of a gonorrhœa,
but only that there are certain remedies which may

advantageously be given through the stomach to ultimately act with benefit upon the diseased urethral mucous membrane.

I shall first consider the treatment of an uncomplicated case of gonorrhœa, reserving that of the complications for subsequent discussion. You will please remember that gonorrhœa is a catarrhal affection of a certain mucous membrane, and that it is entirely local. In its very earliest stages, before the discharge has become purulent, a treatment called the abortive is advised in many treatises on Venereal. This consists in adding to the already existing inflammation a more severe one, in the hopes that the greater will remove the less, which is done by injecting a strong solution of nitrate of silver into the urethra, limiting its action to that portion which is already diseased. I beg you to have nothing to do with it, as, from repeated trials of this method, I am convinced that it is uncertain, is liable to produce very acute inflammation, and oftentimes serious hemorrhage, without retarding or checking the course of the disease. There is no royal road to curing a clap any more than there is to the acquisition of knowledge, and, in the treatment of clap, take as your motto, " festina lente."

The best injection is either the sulphate or acetate of zinc dissolved in water ; but before using this let me explain to you the stage of the gonorrhœa in which you will find it of most service. During the acute inflammatory stage, when febrile symptoms are present, when the penis is hot, inflamed, and œdematous, when the mucous membrane of the parts is congested, and there is eversion of the lips of the meatus, with a scanty mucous or muco-purulent discharge, your first object should be to relieve these symptoms, and the use of injections in this stage is entirely inadmissible. For the relief of the febrile symptoms I know of nothing which will take the place of aconite, in small doses frequently repeated, thus :

℞. Aconit. radici tinct.............. ℳi.–ij.
Sig.—In a little water every hour.

To relieve the œdema and swelling of the penis use cold-water dressings, or wrap the organ up in a cloth wet with the lead and opium wash, which is administered as follows :

℞. Liq. plumb. subacetat.,
Tinct. opii.............āā ℥ i.
Aquæ .... q. s. ad ℥ viij.
M.
Sig.—Locally.

The diet during this stage should be of the light-est, such as milk, milk-porridge, gruel, and the fari-naceous articles of food. These symptoms usually disappear in the course of forty-eight to seventy-two hours, when the discharge becomes purulent and abundant, and often associated with a frequent desire to pass water, not from any invasion of the neck of the bladder, but simply from reflex action, due to the local irritation within the first inch of the urethra. Now is the time to begin with injections, and of these the best, as I have already said, are the preparations of zinc :

R. Zinc sulph.................. gr. viij.–xij.
Aquæ...................... ℥ iv.
M.
Sig.—To be injected thrice daily.
Or—
R. Zinc acet.................... gr. viij.–xij.
Aquæ...................... ℥ iv.
M.
Sig.—Inject thrice daily.

Alum, either alone or in combination with tannin, as well as tannin alone, have been advised as injections, but, in my estimation, they possess no advantages over the preparations of zinc. They may be used as follows :

R. Alumin. sulph.................. gr. xx.
Aquæ .... ℥ iv.
M.
Sig.—As injection, thrice daily.

℞.   Alumin. sulph.,
Acid. tannic. pulv.............āā gr. x.–xv.
Aquæ......................... ℥ iv.
M.
To be well shaken before using.
Sig.—Inject thrice daily.

One of the objections to the use of tannin is the persistent stain which it leaves upon the body-linen, but I shall shortly mention a simple manner of obviating this.

An injection which is known as Ricord's formula is often used, and is an excellent one. It is composed of the following ingredients :

℞.   Zinc. sulph...................... gr. viij.
Plumb. acet..................... gr. xv.
Tinct. opii.,
Tinct. catechu...............āā ʒ ij.
Aquæ.......................ad ℥ iv.
M.
Sig.—As injection, thrice daily.

To obviate the staining of the clothes, either from the disease or from the injections used, a false front may be made by pinning to the shirt a double fold of unbleached cotton the size of the front flap. It also has the advantage of keeping the parts clean and cool. Never countenance wrapping up the penis in innumerable folds of linen or cotton, which

is so often done, as it keeps the parts in a heated
state, prevents the free exit of the pus, which, from
its irritation, is very prone to produce balanitis and
œdema of the prepuce.

As regards injections, there are some points to
which I wish to call your attention, for upon the
proper employment of this class of remedies will
often depend the efficacy of treatment.   In the first
place, never use a glass syringe if you can help it ;
the fluid nearly always comes out behind the piston
instead of through the nozzle, and the patient re-
ceives little, if any, of the injection.

The syringes made of vulcanized rubber are the
only ones which are fit to be used, but even some of
these are objectionable from their inordinately long
nozzle, which, when inserted within the meatus, ·
throw the injection beyond the seat of disease.   No
urethral syringe, designed for use during the earlier
stages of clap, should have anything but a very
short point, and the best are those which terminate
in a cone.   See that the piston works easily and
readily, without any jerking movement, and that it
admits of no leaking behind ; also be careful that
the instrument is sufficiently large to hold a couple
of drachms of fluid.   The little pocket-syringes which
are sold in shops are catch-penny affairs.

Now as to using it.   The syringe being carefully
charged with the injection, and all air excluded from
the barrel, the patient holds the instrument in his

right hand, between the thumb and second finger, the index-finger being stationed at the butt-end of the piston. The penis is held between the second and third fingers of the left hand, the palm looking upward, the index finger and thumb being left free to separate the lips of the urethra. The nozzle of the syringe is then carefully inserted just within the meatus, when the end of the urethra is closed against the instrument by a gentle lateral pressure. Do *not* place the finger and thumb above and below the meatus, otherwise you will open the canal instead of closing it, and the fluid will escape as fast as it is injected ; but make a gentle pressure sideways, and if this be properly done, none of the fluid will run out. Now with the right hand gently drive the piston home, without any sudden movement, and if the syringe is in proper working order, this is readily accomplished. As soon as this is done, and all the fluid deposited within the urethra, with a quick movement withdraw the nozzle of the syringe from the urethra with the right hand, while with the thumb and index finger of the right hand still in position, the patient closes the meatus. This prevents the outflow of the injection. Then laying the syringe down, the patient with his right hand gently strokes the floor of the urethra from behind forwards in order to press the fluid as far as possible into the anterior portion of the canal, which is the seat of the disease in the earlier stage. As the disease invades deeper parts, this motion must

8

be reversed in order to crowd the fluid backwards. After the injection has been retained from two to five minutes, the compression with the left hand upon the lips of the meatus may be discontinued, when a por-tion of the fluid will run out. The injection should cause a slight sensation of warmth and tingling in the canal for five or ten minutes after its use, but this should never amount to actual pain; if it does, it shows that the injection is too strong, and it must be diluted.

Before using the injection, it is a good plan to make the patient pass his water, in order to wash out any of the discharge which may still be lodged in the urethra, and should this be impossible, cleanse the canal with a syringeful of tepid water before using the medication.

Within a few years a method of treating gonor-rhœa by means of medicated bougies has been advo-cated. They are made of coca butter, holding an astringent in minute subdivision, and are left within the urethra to melt. Experience has not shown me that they have any special advantage over injections, and they have the decided disadvantage of being dirty and disagreeable.

Internal treatment consists in the use of those rem-edies which are excreted by the kidneys, and which contain a balsam or resin; foremost among these are copaiba, cubebs, and the oil of yellow sandal-wood. In order to cover their nauseating taste, they are

given either in pill form or in capsule, in the follow·
ing manner :

R. Copaibæ..........................℥ i.
Oleo-resin cubebæ.................℥ ss.
Magnesiæ.........................q. s.
M. Ut ft. massa.
Divide into pills of five grains each.
Sig.—Three to six three times daily, after meals.

If given in capsule, the balsam of copaiba, the oil
of cubebs, or the oil of sandal-wood is employed, each
capsule being supposed to hold ten minims of these
various drugs.

This is by far the neatest and best way of adminis-
tering these drugs, as, when given in solution, they
are extremely nauseating, and are apt to produce
vomiting, and the medicine has to be suspended.
If used in a fluid form, what is known as Lafayette's
mixture is the least objectionable. The following
is its composition :

R. Copaibæ...................... ℥ i.
Liquoris potassæ.............. ℥ ij.
Ext. glycyrrhizæ............... ℥ ss.
Spiritus ætheris nitrici.......... ℥ i.
Syrupi acaciæ................. ℥ vi.
Olei gaultheriæ............ ... gtt. xvi.
M.

The copaiba and the potassa should be first mixed to·
gether, the liquorice and nitre added separately, and the

remaining ingredients together.   They should then be
thoroughly incorporated and given in tablespoonful doses
p. r. n.

These preparations have the effect of relieving the
ardor urinæ, and of checking the discharge.   They
should always be given after meals, as they are then
less liable to disturb the stomach or to produce
nausea.

Powdered cubebs is sometimes given in the early
stages of a clap, with the view of relieving the ardor
urinæ, and is best administered in the form of a
wafer, which is easily swallowed and prevents the
drug from being tasted.   Its action seems to be prin-
cipally confined to rendering the passage of the
urine easy, as it has little effect upon the discharge.*

Another remedy of recent use is the kava-kava,
or piper methisticum, a root the juice of which is,
or used to be, of common use in the islands of the
South Pacific Ocean for purposes of stimulation.   It
is generally given in the form of a fluid extract, in
doses of thirty minims to a drachm, several times
daily.   I have not used it sufficiently to enable me
to decide upon its actual value in the treatment of
these diseases, and I cannot do further than mention
it as one of the numerous promising remedies ad-

---

* A very simple, at the same time effective, way of relieving the ardor
urinæ is to make the patient pass his urine in a tumbler nearly full of
hot water ; in other words, pass his water under water.

vised in gonorrhœa. As regards the others I have named, I can speak positively, and can particularly commend the oil of the yellow sandal-wood. Two objections may be urged against it : first, the difficulty of getting it pure ; and second, its expense. If, from either of these two causes, it should not be given, the copaiba is, to my mind, the next best drug given in pill or capsule, as already noted.

The stage at which these remedies should be administered is where the disease begins to show signs of subsidence, although they may often be used when there is much pain during micturition, and a copious discharge. If the patient's stomach will bear them, the effect is sometimes wonderful ; but their long continuance is liable to induce pain in the region of the kidneys, and a deposit in the urine which has been mistaken for albumen.

As regards diet, the rules must be strictly laid down, and no deviation allowed until the disease has entirely disappeared. Except during the acute inflammatory stage, the patient should not be kept upon a low diet, but ought, on the contrary, to be allowed a good and nutritious regimen. Meat, vegetables, fish, eggs, and the like may be allowed, and I beg you to remember that by half-starving your patient you only tend to keep up the gonorrhœa. Asparagus, highly spiced dishes, strong coffee or tea, and, above all things, every form of alcoholic or malt beverage, as well as immoderate smoking of tobacco, should be interdicted in the majority of cases; but if the patient has been accustomed to use them, a weak cup of coffee or tea well diluted with milk may be allowed once daily. Lemonade, the copious use of mineral waters, and cider, should also be tabooed, and the patient confine himself to water or milk as drinks. Flaxseed-tea has been recommended, but it is usually such a nauseous mess that the patient is only too glad to drop it out of his list of beverages. In the summer-time there is no objection to the use of the ripe fruits, and their juice often makes an agreeable addition to water; but such drinks should be sparingly sweetened, as a portion of the sugar is converted into alcohol in its passage through the human system.

The use of injections in the female deserves a few words. The amount injected is much greater than in the male, and is employed in conjunction with other remedies, which should never be used except by the surgeon. The vagina being first cleansed by the use of tampons of prepared cotton, introduced through the speculum, should then be painted over with a strong solution of nitrate of silver (gr. xx.–xl., to aq. $\frac{2}{3}$ i.) or the pure tincture of iodine. If care is taken not to allow the fluid to run out upon the vulva or the external genitals, no pain is felt, as the vagina and the cervix uteri are not sensitive parts; and even if the medication does reach those portions, the smarting is not very severe, nor does it last long. Upon the withdrawal of the speculum a layer of dry cotton is placed between the labia to separate them as well as to prevent the discharge from trickling down over the perineum and the inside of the thighs, and to obviate excoriation of these parts. The patient should then be directed to use one of the following injections in the manner I shall describe to you in a few moments :

    ℞.   Aluminis pulv................... ℨ i.– ℨ ij.
        Aquæ tepidæ................... ℥ viij.–xij.
    Inject thrice daily.

Or—

    ℞.   Ac. tannic.................... ℨ ss.– ℨ i.
        Aquæ tepidæ................... ℥ viij.–xij.
    M.
    Inject thrice daily.

Or these two may be combined, thus :

    ℞.   Alum. pulv.................... ℨ i.– ℨ ij.
        Ac. tannic.................... ℨ ss.– ℨ i.
        Aquæ tepidæ................... ℥ viij.–xij.
    M.
    Inject twice daily.

Remember, however, that the tannic acid stains permanently, and some women object to having this constant reminder of a former clap on their linen.

A very good injection is to add to a half-pint bottle of ordinary table claret—

Alum. pulv.................... ℥ ij.
Zinci sulph.................... ℥ i.

of which the patient is directed to use from one to three tablespoonfuls, in from one-half to a full pint of tepid water, thrice daily.

It is well, for convenience, to give women the materials in bulk and let them mix their injection themselves, reckoning a teaspoonful as the equivalent of the drachm.

If the inflammation is very acute, no injection should be used, except one of *hot* water and the frequent use of *hot* sitz-baths is advisable, but as soon as the symptoms subside, the medicated fluids should be employed.

A very good method of keeping the astringent in constant apposition with the diseased mucous membrane, is by the use of vaginal suppositories, which may be made of the strength of gr. ij.–v. of the astringent to each suppository. A very good substitute is to soak a pledget of prepared cotton in a solution of tannic acid ℈ ij., to glycerine ℥ i., and lay it on the diseased parts. Other astringents may be used in the same manner, such as alum of the same strength as the tannin given above, or the tincture of catechu without the glycerine. Be careful to remove these tampons frequently (three or four times daily), else trouble will ensue from decomposition of the retained discharge, and also remember that in all these diseases cleanliness, if not superior, is next to godliness.

In using injections, in order to make them effective the following rules should be observed. The glass and rubber syringes which are often sold under the name of vaginal syringes, are of no earthly use ; the only effective one is the Davidson's rubber syringe, or, better yet, the vaginal douche invented by Dr. Frank P. Foster, of New York City. In giving an injection, the woman should never be allowed to assume a squatting posture, as the fluid runs out as fast as it is thrown in, and does not reach the deeper portions of the canal ; but she should be placed upon her back, with the hips slightly elevated, when the vagina is by the force of gravity thrown open, and the fluid, by the same physical action, is carried into every portion of the canal. Some of the fluid, of course, escapes · and in order to protect the woman's clothing, a sheet of rubber cloth should be placed under the hips, and a vessel in readiness to catch the overflow. Foster's vaginal douche is better yet, inasmuch as it has an overflow pipe to carry off the superabundant fluid, and the injection can thus be administered with thoroughness and convenience. At the close of the operation, a tampon of dry cotton should be placed between the labia, to retain what fluid is left in the canal. These injections, remember, are to be used in conjunction with the applications

8*

which the surgeon makes himself every second or third day.

The treatment of the complications in gonorrhœa vary according to their character. For *balanitis*, the most important point to be observed is *cleanliness*, and this, in many cases, will be all that is required. In severe cases, upon exposure of the glans penis by retraction of the prepuce, the parts may be painted over with a solution of nitrate of silver, from five to ten grains to the ounce of water, and the subsequent dressings should be either of ordinary starch powder, the impure oxide of zinc (calamine), or lycopodium, and a thin layer of lint or prepared cotton placed between the prepuce and glans penis.

For phimosis, if incomplete, subpreputial injections of warm water, or of a slightly carbolized lotion, with proper attention to cleanliness, will generally be sufficient ; when it is complete, and especially if the foreskin acts as a reservoir for the pus and urine, circumcision should be practised, provided the inflammation is not very acute, and there be no œdema. There is no danger in the operation so far as the clap is concerned, for the secretion, you know, is not auto-inoculable ; but, of course, be careful that the discharge does not come from concealed chancroids instead of gonorrhœa, and auto-inoculation will here give you the requisite information as to its nature.

Paraphimosis, which is the opposite of phimosis, is relieved by compression of the glans penis with the right hand so as to squeeze all the blood from the part; traction forwards of the prepuce is then made, by grasping it posterior to the constricted portion between the fingers and thumb, which are held in the shape of a circle, the penis lying in the enclosed space between the fingers. At the same time that the forward movement is made, the glans is pushed backwards in the hopes of forcing it beneath the constriction. If this be not successful, an incision must be made through the strictured portion of the foreskin, when the prepuce can be drawn forward over the glans, and as soon as the inflammation and thickening of the foreskin has subsided, the unseemly dog's ears which are left behind may be removed by circumcision.

Chordee, of all complications, is the one that will put you to your trumps to relieve. Everything in the pharmacopœia has been tried, and, I might almost say with truth, has been found wanting. Lupulin, camphor, belladonna, opium, bromide of potassium, ice, and hot water have all been used with varying success; but, to my mind, the one remedy which gives the most relief is the hypodermic injection of morphia and atropia,* given in the perineum or the

---

* ℞. Atropiæ...... .... gr. i.
    Acidi acetici.. .... q. s. ut ft. solutio cum,
    Aqua destil... .... ℥ iv.
Et adde,
Magendie's solution of morphia q. s. ad unciam unam.
Of this inject 5–8 minims hypodermically.

inside of the thighs at bed-time. In giving these in-jections in the perineum, you must of course be care-ful not to wound the membranous urethra by carry-ing the needle too deep ; and if you select the inside of the thighs, be careful not to puncture the internal saphenous vein. All of these dangers may be avoided by making your punctures just beneath the skin. Of the internal administration of remedies, camphor and opium, or camphor and belladonna, or opium and belladonna, give the best results, thus :

    ℞.   Pulv. opii........................ gr. i.
         Pulv. camph...................... gr. ij.
         Sacch. alb....................... q. s.
         Ut fiat capsula una.

Sig.—One at bed-time, and repeat in two hours if necessary.

    ℞.   Belladon. extr. alcohol.......... gr. ss.–i.
         Camph. pulv..................... gr. ij.–iv.
         Ut fiat capsula una.

Sig.—One at bed-time, and repeat if necessary.

    ℞.   Pulv. opii....................... gr. i.–ij.
         Extr. belladon. alcohol.......... gr. ss.–i.
         Ut fiat pil. una.

Sig.—At bed-time, and repeat if necessary.

The genital organs may, in addition, be bathed at bed-time in perfectly *hot* water, which will sometimes relieve the tendency towards erection, and the hot I have found of more service than cold applications. A method of immediate relief more generally practised among the lower orders abroad than here, is to place the penis during the state of erection upon a table or flat surface and strike it a smart blow with the fist upon its convex surface. It certainly relieves the chordee at once, but at the expense of profuse hemorrhage and a subsequent traumatic stricture, for the urethra is ruptured by the blow. It is hardly necessary for me to add that I do not advise your practising any such method.

When the *testicles* are affected, the first step in the treatment is to insist upon the patient's going to bed, and the old maxim of Malgaigne in such cases is a good one : " The patient on his back, and his testicles towards the ceiling." If the patient should at first be restive under such advice, you may be very sure that sooner or later he will accede to it, for his testicles will continually remind him that he is a fool to stand when he can lie, and he will perforce be glad to seek his bed to escape the intense suffering which gonorrhœal epididymitis involves. If the inflammation be very acute, leeches should be applied ; but I strongly advise you never to put them upon the scrotum. I know it is done, and sometimes without producing trouble ; but, on the other hand, I have seen an enormous subcellular infiltration of blood take place, which, although not dangerous, is dis·

agreeable, and your patient already suffers sufficient discomfort without your adding to it. Place the leeches, then, at the external abdominal rings, at the perineum, or along the inside of the thighs, and let their number be six or eight ; in other words, *abstract blood freely.* Unless the inflammation be very acute, blood-letting will not be requisite, as other remedies will be fully as serviceable ; and of these, applications of cold are the best. Pack the testicle in ice, which should be finely broken up and placed in a water-tight rubber bag, or, what will answer the same purpose, a well-made rubber condom. This will often relieve the pain and make the patient comparatively comfortable and easy. Poultices of hot flax-seed meal, or of tobacco leaves, soaked in hot water, are sometimes used ; but all these applications are nasty messes, and if you deem heat requisite in the treatment of these diseases, a very good way of applying it is by soaking a preparation known in the shops as spongio-piline in hot water, and wrapping the testicle up in that. Flannels wrung out in hot water will oftentimes serve the same purpose ; but nine times in ten cold applications answer better than hot ones. The ice should be steadily persevered in until the pain is relieved, unless, indeed, its application causes discomfort to the patient from too great a degree of cold, when its use may be intermitted. The testicle, in the meantime, should be well supported and not allowed to hang between the patient's thighs. A very neat manner of relieving the pain is by making multiple punctures with a surgeon's needle, the larger the better, or with a bistoury, which is guarded to within a quarter of an inch of its point. The testis is grasped in the left hand and several rapid punctures are made into the swollen epididymis, care being taken not to make them too deep. Blood and serum follow the punctures, and oftentimes immediate relief is experienced.

After a week or ten days the pain in the testis subsides, and the patient is able to leave his bed, when upon examination the epididymis is found enormously enlarged and indurated, and still tender upon pressure. This swelling gradually subsides, and under very favorable circumstances may entirely disappear, but, as I already stated in the last lecture, some thickening is nearly always left behind. To reduce this, an ointment composed of equal parts of unguent. hydrargyri and unguent. belladonnæ may be applied to the testicles upon a piece of linen or soft kid. Another plan, but one little used at the present day, is strapping the testicle in the manner described in your manuals on surgery; but as this has sometimes led to atrophy of the testis, its general use has been pretty nearly abandoned. One remedy which has been lately advocated is the internal administration of tincture of pulsatilla, which is used in doses of one minim, or fractions of a minim, repeated every hour for twenty-four to forty-eight hours. In the cases in which I have used it, I have been rather pleased with the results, as it has seemed to have the effect of relieving the pain rapidly, although it has no effect upon the swelling. The preparation which I have used has been the homœopathic mother tincture, which is better prepared, and more certain in its results, than the non-officinal preparations which are sold by the druggists. To relieve the enlargement of the epididymis, iodide of potassium in 5 or 10 grain doses, three times daily, is sometimes advised, but my experience has been that this salt makes the clap worse. I therefore have given up its use, and substitute for it the simple tincture of iodine, in 5 or 10 minim doses, which I do not find open to the same objection as the iodide.

The treatment of *prostatitis* is twofold, the first being directed to the relief of the inflammation of the prostatic urethra, and the second to the prevention of suppuration in the swollen organ. The symptoms, you remember, of prostatitis and cystitis were very similar; hence, what will answer for one will also do for the other. If the prostate, upon rectal examination, be found very much enlarged and tender, leeches may be applied to the perineum, and as soon as they have come off, hot fomentations should be applied. For the relief of the dysuria, rectal suppositories of opium and belladonna should be used, of sufficient strength to insure freedom from pain. They should be prepared as follows:

℞. Extr. opii........................ gr. ij.
Extr. belladon................... gr. l.
Theobroma .... q. s.
M.
Ut fiat suppos. No. 1.
Sig. p. r. n.

Pieces of ice may also be passed up the rectum, and kept in apposition to the inflamed prostate.

Injections in both prostatitis and cystitis should be suspended, nor should they be resumed until the acute inflammation has entirely passed away.

The internal remedies which have been advised for these complications of gonorrhœa are numerous, but, to my mind, four-fifths of them are useless, and had better be dropped out of the list. The best are copaiba, or the oil of yellow sandal-wood, and, if a diluent is required, sweet spirits of nitre, well diluted with water, given several times during the day in table spoonful doses. The homœopaths frequently use minute doses of the tincture of cantharides, in fractions of a minim to one or three minims every hour ; and, in some cases, where I have used the fly in these minute doses, I have given relief when the neck of the bladder was implicated ; in other cases it has entirely failed. When the tincture of cantharides is used, the preparation should be fresh.

When suppuration threatens, the formation of pus should be favored as much as possible by the use of hot sitz-baths and hot fomentations to the perineum, and the surgeon's efforts should be directed to make the abscess point into the rectum. As soon as fluctuation is felt, open it and dress the part afterward with injections of warm water, to which may be added a little carbolic or nitric acid, but these should be very weak, the principal object being to keep the abscess clean and free from the accumulation of pus or fæcal matter. The cut edges of the wound will often form a sort of valve, which acts as a protection against the retention of foreign bodies in the abscess.

During the acute inflammation of both these organs the gonorrhœal discharge, almost if not entirely disappears, to return again as soon as it has passed off, and when this occurs, injections can again be used. The patient should now be instructed, after throwing in the injection, to work it back as far as possible into the canal, and this may be done by stroking the urethra from before backward, in order to press the fluid into the deeper parts of the canal. The surgeon may himself, two or three times a week, make a deep injection with a long-nozzled urethral syringe, of one of the following preparations :

R. Argent. nitrat.................. gr. ss.–i.
    Aquæ........................ ℥ vi.
M.
Sig.—For deep injections.

℞. Tinct. ferri persulphat.......... ℨ ss.– ℥ i.
    Aquæ...................... ℥ vi.–viij.
M.
Sig.—For local use.

These injections should never be intrusted to the patient, but the surgeon should always give them himself.

For *cowperitis* and inflammation of the vesiculæ there is little to be done beyond rest, the application of leeches to the perineum, and the use of the rectal suppository of belladonna and opium. If suppuration threatens, favor it as far as possible, and as soon as the abscess is ripe, open and treat it as you would abscesses elsewhere.

After the discharge has lost its purulent character and subsided into the condition known as gleet, the treatment undergoes certain modifications. If the gleet be dependent upon a stricture, this must be removed before the gleet can be cured, which may be done either by gradual dilatation with bougies and sounds, or else by one of the many operations advised in your text-books on surgery. When it is not dependent upon gleet, but is due to a subacute inflammation of the deeper portions of the urethra, the stronger medicated injections should be stopped at once, and cold water or very weak, astringent solutions thrown into the canal several times daily. The surgeon should, twice or three times weekly, pass a steel sound of the largest size the urethra is capable of receiving, which should be withdrawn within a few seconds after its introduction into the bladder. A steady perseverance in this course of treatment for a few weeks will generally bring about a cure, and it may be hastened in some instances by the internal administration of the balsamic and resinous preparations of which I have already spoken.

This brings me to one point of my subject which it is well for you to remember : *a discharge is sometimes kept up by over-medication.* Patients will apply to you with the following history : they have been under treatment two or three months for a gonorrhœa, which, after running through its usual course, has ended in a thin, mucous discharge, perhaps only apparent in the morning, and occasionally during the day. There is no irritation while passing water, and, but for the slight discharge, they would be entirely well. This, however, has persisted for several weeks without any apparent change, and has been a source of worry and anxiety. The patients have lost flesh and strength, while the face will often bear signs of the mental excitement under which they are laboring. Bid such patients throw away their syringes, stop all injections and clap medicines ; bid them live well, and use with their dinner a moderate quantity of some light wine—the red Bordeaux wines are the best; advise them against beer and spirits at the start, but these may be used later on if deemed requisite. Tell them plainly that they are keeping up the discharge by over-treatment, and that the sooner they can recognize the fact the quicker they will get well. Sometimes nothing further will be needed, but occasionally some tonic, such as the tincture of the chloride of iron, or the syrup of iodide of iron, in doses of 5 to 15 minims, may be given with advantage, and you will have the gratification of hearing the patients tell you in a short time that they are entirely well.

I am bound to say, however, that some cases of gleet will last for years in spite of all kinds of treatment; but there is one consolation that you can afford your patient, sorry though it may be, that some time or another it will come to an end, and I have seen such cases recover, apparently from nothing else than a change of climate, or a sea voyage. These cases, however, are comparatively few.

The treatment of gonorrhœal complications in the female deserves separate consideration. In *vulvitis*, one of the first things to impress upon the patient's mind is attention to cleanliness, inasmuch as the discharge from the inflamed parts tends to keep up the cause which gave rise to it, and also to produce irritation of neighboring parts. The surgeon should make an application every two or three days of a weak solution of nitrate of silver, one to three grains to the fluid ounce of water, and the patient directed to place pledgets of lint or prepared cotton between the inflamed labia, the lint being previously soaked in solutions of alum or the liquor plumbi subacetatis. They may be prepared as follows :

    ℞. Pulv. alum.................... gr. x.-xv.

        Aquæ....................... ℥ i.

    M.

        For local use. ·

Or—

    ℞. Liquor. plumb. subacetatis, one part.

        Aquæ, two parts.

    M.

        For local use.

Such cases are particularly adapted, during the acute stage, for the use of the dry dressings either of lycopodium, of calamine, or finely pulverized starch. As soon as the acute inflammation passes off, the astringent applications, advised above, should be made.

The inflammation of the vulvo-vaginal gland may at first be treated in the hope of preventing the formation of an abscess, and this is done by painting the part with a ten-grain solution of nitrate of silver, and by the topical application of equal parts of belladonna and mercurial ointment applied on a piece of linen or cotton. If suppuration be inevitable, hot sitz-baths and poultices should be used to favor the formation of matter, and the resulting abscess opened as soon as fluctuation is detected, provided the pus cannot be evacuated through its natural duct.

For *urethritis* there is nothing better than the use of the solid stick of nitrate of silver passed over the entire surface of the urethral mucous membrane, which is easily accomplished, owing to the shortness and peculiar situation of the urethra in the female. The pain of the application is not as great as you might imagine, and speedily passes away. It is here that the internal administration of copaiba, cubebs, and sandal-wood oil are of use during a gonorrhœa in women ; in the other varieties of the disease they exercise no effect.

In *inflammation of the cervix*, the part should first be thoroughly cleansed from all discharge with a piece of cotton wound on the end of a uterine probe, and the canal touched with a solid stick of the nitrate of silver, care being exercised that the cervix alone, and not the body of the uterus, is cauterized. The patient should not be trusted to make any applica-tions herself, owing to the danger of exciting inflam-mation in the body of the womb, the treatment of this portion of the genital apparatus being left entirely in the surgeon's hands. When the *body of the uterus* and its appendages are attacked by gonorrhœa, the results are sometimes very serious ; peritonitis and pelvic cellulitis resulting from the discharge flowing into the pelvic cavity. The patient should be at once confined to bed, and local abstraction of blood by leeches resorted to. This should be followed by the application of hot fomentations to the uterine re-gion, and by the internal administration of opiates in sufficient quantity to keep the patient free from pain. If all goes favorably, the discharge will be evacuated through the cervix ; under other circumstances, pel-vic abscess will ensue, which will be evacuated some-times through the vagina, sometimes through the rectum, and sometimes through the abdominal walls. Gonorrhœal endometritis is always a serious complica-tion, which fortunately is not of frequent occurrence.

When the *ovary* is inflamed, rest in bed, leeches, hot fomentations, and painting over the inflamed spot with the compound tincture of iodine, are the best means to be employed.

The treatment of *gonorrhœal rheumatism* is as un-satisfactory as it can well be, for there is no form of rheumatism which is more rebellious to the action of remedies than this. The ordinary internal remedies are of no avail, and those which promise the most success are the local application of blisters above and below the affected joints, painting the affected parts with the compound tincture of iodine, and the inter-nal administration of the iodide of potassium. But even these sometimes prove of no service, and the case goes on to anchylosis, partial or complete. In cases where pericarditis ensues, the local applications of the strong tincture of iodine should be made over the pericardial region, and the iodide of potassium in five to ten grain doses administered internally three times daily. But sometimes permanent thickening of the cardiac valves takes place, just as it does in rheumatic pericarditis from other causes.

Gonorrheal ophthalmia is of importance, according to the form which it takes, and the treatment varies widely. When due to contagion from the conveyal of matter by the fingers, the attack is extremely serious, as the eyeball may be destroyed within forty-eight hours, unless prompt measures be taken for its relief. The eyelids and the eye itself should be kept sedulously and carefully clean by frequent syringing with warm water every ten or fifteen minutes; the eyelids should then be everted, so far as the enormous œdema and swelling will permit, and the parts brushed over with a strong solution of nitrate of silver of the strength of 40 to 60 grains to the fluid ounce of water. Where the œdema and chemosis is very great, blood should be abstracted from the temple by leeches, or by the instrument known as Heurteloup's artificial leech, or by incisions into the swollen mucous membrane. *Remember, never to put the leeches on the eyelids.* In spite of all care and attention, corneal ulceration will sometimes go on very rapidly, and the contents of the eyeball be evacuated. The subsequent thickening and granular condition of the lids, as well as the keratitis and chemosis, should be treated by the methods laid down in the text-books on ophthalmic surgery.

The other form of gonorrheal ophthalmia is not so serious. The conjunctivitis and the serous iritis may be treated in the case of the former by repeated bathing with warm water, blisters to the temples, and the instillation of the following collyrium :

R. Soda bicarbonat................... gr. x.
    Aquæ camph..................... ℥ ij.
M.
Sig. p. r. n.

The serous iritis may be treated by dropping into the eye, three or four times daily, the sulphate of atropia, four grains to the ounce of water. If the iris remains sluggish to the action of the atropine, one or two leeches may be applied to the temple or over the supraorbital region.

Inflammation of the lymphatic inguinal glands, or those running over the dorsum penis, should be treated in the earlier stage by rest, leeches, blisters, the daily application of the tincture of iodine, and the internal administration of the sulphide of calcium, $\frac{1}{10}$ to $\frac{1}{5}$ of a grain, three times daily.

Should, however, these measures prove ineffective to prevent suppuration, it should be favored as far as possible by the application of poultices, and as soon as fluctuation is detected the bubo should be opened in the method laid down in the chapter on chancroidal buboes, care being taken that all sinuses are freely laid open whenever and wherever they present themselves. The pus of such buboes is always laudable and never contagious; indeed, they are nothing more than abscesses of glandular or periglandular tissues. Should the gland itself suppurate, recovery may be hastened by its removal either with the knife or ligature.

Warts, or the broad condylomata, as they are sometimes called, if small and pedunculated, may be snipped off with scissors and their bases touched with strong nitric or acetic acid. When they are large, or seated upon a broad base, they should be painted with strong acetic acid, and dusted over with powdered alum, or with the dried sulphate of iron, mixed with equal parts of lycopodium. When very exuberant and large, especially in those cases which occur in both sexes on the nates and perineum, I have often injected two or three minims of glacial acetic acid into the substance of the wart with benefit. It shrivels up the growth with surprising rapidity, and, if properly used, injecting but a few drops at a time, is not attended with any danger; at the most, an abscess is the worst result that will follow, unless, of course, the acid is used recklessly and beyond the bounds of prudence. But one point I wish particularly to impress upon your minds in the treatment of these affections: *keep the parts dry and clean ; it is four-fifths of the treatment.*

Herpes, when slight, is best treated by powdering

the parts with calomel, lycopodium, calomine, or some such dressing ; if they show any tendency to ulcerate, touch them lightly with the solid nitrate of silver, and finish the treatment with the dry dressing above advised. *Do not use wet dressings ;* they only serve to macerate the epithelium and keep the parts in a condition of moisture unfavorable to recovery.

Where complicated with digestive troubles, these latter must be treated by the remedies applicable to such diseases.

www.ingramcontent.com/pod-product-compliance
Lightning Source LLC
Chambersburg PA
CBHW021708210326
41599CB00013B/1570